MANAGEMENT IN CIVIL ENGINEERING

MANAGEMENT IN CIVIL ENGINEERING

A Practical Handbook

Edited by

PETER C. G. ISAAC

Professor of Civil and Public Health Engineering
University of Newcastle upon Tyne

ORIEL PRESS

First Published 1971
ISBN 0 85362 098 9
Library of Congress Catalogue Card Number 78-127063.

All enquiries relating to any of the articles appearing in this book should be addressed to the publishers.

Published by Oriel Press Limited
32 Ridley Place, Newcastle upon Tyne
England NE1 8LH

Text set in 11 on 13 point Baskerville

Printed in Great Britain by Bell and Bain Ltd., Glasgow

CONTENTS

PREFACE

At the end of September 1970 Professor W. Fisher Cassie CBE retired after thirty seminal years as head of the Department of Civil Engineering at the University of Newcastle upon Tyne. When I and my colleagues decided to honour him with a symposium in September 1970 it was clear that, since he had made his mark on so many branches of civil engineering, it would be appropriate only if we chose a topic of general and fundamental importance to the whole profession. I was reminded of Thomas Tredgold's definition of civil engineering as the 'art of *directing* the great sources of power in nature for the use and convenience of man', and it seemed to me that the increasing emphasis in recent years on ever more sophisticated design techniques had diverted attention from the fundamental importance of *directing* to the practice of the profession. We decided, therefore, to devote the symposium to a discussion of management in civil engineering, directing attention particularly to those aspects which can by discussed objectively. The result was the collection of papers here presented; they provide such a valuable and synoptic review of management in civil engineering that it has been decided to publish them as a text, completed by Mr. Maurice Milne's summary of the two days of discussion and Dr. Wearne's bibliography. The full edited discussion on the papers is also being published separately for those who attended the symposium. It is believed that this book will be of value to all those in the profession who have any kind of wider responsibility or who aspire to such responsibility. It is hoped that it may also help those young civil engineers who have to sit for the paper on the Engineer in Society of the Council of Engineering Institutions.

My especial thanks are due to all the authors who bore most graciously my many importunities, to the small Papers Committee organized under the aegis of the Institution of Civil Engineers, namely Messrs. T. N. W. Akroyd, H. D. Manning, M. Milne, and F. H. Woodrow and Capt. A. G. Reid, to Dr. S. H. Wearne, and to the many others who assisted in so many ways to organize the Symposium and to produce this book.

Newcastle upon Tyne   Peter C. G. Isaac
March 1971

MANAGEMENT IN CIVIL ENGINEERING

S. Erskine-Murray

IT SEEMS appropriate to begin with a consideration of what we mean by the term management. Only then can we proceed to discuss the skills appropriate for management, the personnel who require to be proficient in them, and how and when these skills should be acquired.

I believe that everyone has a different view of what management means and this impression is perhaps supported by the multiplicity of publications that deal with the subject. There are in fact over 1000 books in the library of the B.I.M. having the word management in their titles, but my own experience is that the majority of them, if studied, would give little help in defining the term. The word management has an agricultural origin and has been used to describe the proper husbanding of land. It is easy to understand how the term has been extended to embrace other activities including, of course, all forms of engineering and commercial activity. The best definition I have been able to devise is as follows :

> The direction of men, finance, equipment, materials, technical skills and other resources towards a chosen goal.

This definition can be faulted since goals do not necessarily remain fixed, and the senior management of any enterprise must constantly be reviewing existing, and selecting new, objectives. Assuming, however, that with this exception the definition is acceptable and certainly applies to the majority of civil engineering situations, it follows that few people are not involved in management. Those who lead projects have more resources at their disposal, and therefore require more techniques to exploit them, but a man must be lowly indeed who has not some resources to manage even if they are limited to his own hands. Accepting that all civil engineers need to manage in some degree, what subjects should he study and what techniques will he require to do so effectively?

There are some who say that skills in management come from experience or are best acquired as the need arises. I disagree. There are many subjects a theoretical knowledge of which is just as important as of subjects directly connected with engineering. I will outline a few. Not everyone will agree that they all have a bearing on management, but I sincerely believe that they have.

First I would name the study of human behaviour. The greatest deficiency in the average man is a lack of understanding of the way in which man reacts with man. We have so often witnessed misunderstandings in public life, private life and in business due largely to unexpected and what seems like the irrational behaviour of others. I know that all cannot be perfect in an imperfect world, but a grounding in psychology and sociology can improve matters in many ways. It can help to predict personal reactions, to improve on the selection of staff, to conclude negotiations successfully, and to obtain more readily the cooperation of those with whom we work. The social sciences could well form part of a university engineering course and could be continued by extramural study possibly through programmed instruction or occasional refresher courses.

The second subject that I believe helpful to the engineer, and certainly part of management techniques, is a study of organization, administration and communication. Those of us who have worked in a badly organized firm will know what I mean. No proper definition of responsibilities, which results in the negation of job satisfaction, lack of rapid decisions through attempted detail control from the centre, or alternatively a situation where the left hand is unaware of what the right hand is doing. One can list common deficiencies which could largely be overcome by basic knowledge of the principles involved. It would be interesting to learn the views of others as to how to acquire sufficient knowledge of this subject and thus avoid the common errors that abound. Perhaps this is a subject best dealt with by postgraduate case-history study.

Next I would certainly advocate training in general financial management. This is a necessity to any man who is to extend his career outside purely theoretical design and the junior activities of engineering construction. In my early days, and too long thereafter, I had only the haziest notion of how an engineering project was financed, the principles of cash flow, how to tell from balance sheet and profit and loss accounts whether a company was thriving or in difficulties, and other financial mysteries. Nor was I conversant with the various methods of cost accounting and cost control. There now exists a comprehensive series of programmed instruction manuals on the major part of this subject. I believe they could be extended specifically for civil engineers and used as a basis of a subject within a civil engineer's degree course.

Continuing to deal with the subjects related to engineering management, I have not yet touched on the desirability of formal education in those technical subjects which though not pure engineering, are nevertheless essential if engineering projects are to be carried out effectively. I refer to mathematically based techniques, many of which are included in the term operations research—theory of queueing, theory of probabilities, forecasting, critical-path analysis, and so forth. Subjects like these can be self-taught or be assisted by the many courses now available, or form part of a university or institution curriculum. I have an open mind as to which method is best, but I am certain that they must be part of the armoury of the engineer, and from their nature have to be originally understood at an early stage of a career. The mature can appreciate new ideas and techniques but it is generally the young who best perform at operational level.

The new polytechnics are already introducing these subjects into their industrial engineering courses, and it seems inevitable that their graduates will become better equipped than the professional civil engineer to control major projects at home and overseas, unless civil engineers have an equally broad base. Perhaps a recognized necessity for postgraduate training is the answer, not necessarily following immediately after the completion of a degree course. The management schools set up in London, Manchester and elsewhere do provide this sort of postgraduate training, but the numbers which pass through these establishments must be small in relation to the numbers that could benefit from participation.

It is a short step from mathematically based subjects to the computer, but this is an area which must be understood and knowledge of which must be constantly kept up to date. I welcome the practice of schools, polytechnics and universities in involving their students as a matter of routine in the everyday use of the computer to solve problems by this means. The possibilities of computer calculation must be appreciated by all who wish to take advantage of technological change, and incidentally make use of the very techniques which I touch on in my preceding paragraphs. But to those in a position to use the potential of the computer initiation is not enough. Technical innovations are occurring so rapidly that frequent injections of new thought are necessary to ensure that knowledge gained is not outdated. There are various ways of achieving this end— refresher courses, seminars or assiduous reading of appropriate journals.

A man is better qualified to work with others or control others if he has a broad base of education. Apart from the greater pleasure in living that even a superficial knowledge of the arts ensures, a common base of education is itself helpful in understanding other people's interests and reactions. I believe that Keele University has a common first-year course for all its students and this certainly appeals to me. I have all my life been conscious of the narrowness of my university education. Perhaps things are different at Newcastle, for example, and my comments are thus outdated, but as an example of what I mean the fourth year of my studies was confined entirely to civil engineering subjects with the exception of one period a week devoted to a subject on which there was no examination at the end of the year. This hardly seems the way to create men who not only can lead their own profession, but who will be capable of leading the commanding heights of industry, commerce or public life. I believe that civil engineering is an ideal initiation into the major problems of management and that the shortage of engineers in commerce and public life is most regrettable. One explanation of this deficiency is that engineers are so wedded to their profession that they have no desire to range elsewhere. Perhaps this is so, but could it not be that the majority are not too well equipped to do so either?

I have touched on a number of subjects which should, I believe, form part of the knowledge of the civil engineer. I could of course increase the number, for example, by suggesting that foreign languages are in many cases desirable. Once again new courses in industrial engineering are giving a lead with modern methods of teaching by language laboratories, and a foreign language is in some cases a compulsory subject in a degree course.

For the civil engineering profession the first need is to recognize a case for greater study of management subjects. Perhaps too many senior engineers take the view that young engineers should concentrate on the job of civil engineering design and not fill their minds with distracting ideas on management, a subject which only their elders and betters have any need to practise! I suspect that reasons for this point of view can be found in the deficiencies in management techniques of its advocates.

Once the need is generally recognized then time must be found to fulfil it! One obvious requirement is an extension of the university undergraduate course. This has already occurred in other countries and it seems inevitable that the same happens here. Then there must be a systematic continuation of management training. I say systematic, since the plethora of courses that are now available makes it difficult to choose a rounded staff educational programme. It follows that the painful decision must be taken to make available time for staff to attend the appropriate seminars, courses or lectures.

Finally, there is the essential need for senior management to be themselves knowledgeable in modern management techniques. Junior staff can be sorely frustrated and disheartened if they find that they are not able to practise the new techniques they have acquired, through the ignorance or apathy of their superiors.

This symposium for senior management touching on so many of the subjects that I have superficially covered seems an excellent method of avoiding this possibility. The support that it has evinced indicates that there are many who do appreciate the need to widen the knowledge of the civil engineer and I for one look forward with great interest to the discussions which will ensue during the two days that we are together.

2

ON MANAGING AND BEING MANAGED

R. L. Wilson

MANAGEMENT IS A HUMAN PROCESS

The management process, whether in civil engineering or any other activity, involves the direction of people. The principles and practice of this aspect of management in a professional environment apply not only to managing directors, senior partners or heads of departments, but also to the whole range of middle management and even to the junior member of the organization, who, for a time, has to direct the work of another.

The aims of management can be achieved only if there is willing cooperation amongst the persons associated with the enterprise. Given that an operation has been adequately staffed with trained personnel, the objective of the manager is to ensure that those whom he is directing are motivated to give of their best. The successful achievement of this objective depends greatly on the quality of the manager and the personal relationship between him and those he is managing, and on methods of management based on an understanding of those aspects which affect the behaviour of a person working. The human factor plays a very big part all through the exercise of management competence.

MOTIVATION FOR PEAK PERFORMANCE

'What makes the worker want to work' is a question which has often been posed ever since human relationships in industry have been studied. In a modern and prosperous society, fear of

B

losing one's job or of disciplinary action has little influence on people's behaviour, and so an authoritarian approach to management requiring subordinates 'to work because they have to' is unlikely to have much success. Discipline is the acceptance of the necessary rules of an ordered enterprise; it does not lead to, but results from, high morale; special provisions for the 'maintenance of discipline' are an indication that morale is not adequate. It is therefore generally recognized that a manager should lead rather than drive his subordinates to achieve peak performance. In assessing the nature of leadership, DALE and MICHELON (1966) identified three factors which motivated people in their work :

1. money motive,
2. sense of belonging, and
3. achievement motive.

Adequate reward for the job done and a fair remuneration relative to the job done by others, both within the organization and outside it, are obvious and essential motivators, but are now recognized as being insufficient in themselves. A sense of belonging or involvement can be fostered in a subordinate by a manager taking a personal interest in him, recognizing the value of his work and keeping him informed as to how his contribution fits into the endeavour of the whole organization. To satisfy the first two factors can lead to subordinates having a high morale and being completely satisfied with their jobs. However, DRUCKER (1955) has attacked 'employee satisfaction' as an almost meaningless concept. He argues that it is impossible to differentiate between job satisfaction that is fulfilment and satisfaction that is apathy.

This leads to the importance of the achievement motive put forward by HERZBERG (1959) who feels that the real motivators are sense of achievement, interesting work, and the feeling that the accumulation of achievement will lead to personal growth and recognition. This is, in effect, management by objective whereby a manager delegates to a subordinate the responsibility (and authority) to achieve an objective; the externally imposed spur of fear is replaced by an internal self-motivation for performance. 'To perform one has to take responsibility for one's own actions and their impact . . . one has, in fact, to be dissatisfied to want to do better' (DRUCKER, 1955). Management therefore becomes less a process of driving subordinates towards objectives, or even leading them, but of providing them with an environment in which they can drive themselves.

TOWARDS BETTER MANAGEMENT PRACTICE

With some knowledge of the factors which influence motivation, guidance can be suggested for better methods in the management of people.

(a) To provide reward

The assessment of the proper reward for the job, incentive payments to encourage extra effort and bonuses for good performance is a separate subject on its own. It is not possible to deal with this aspect in a general paper on management. However, although the provision of reward contributes towards motivation, a manager cannot buy loyalty or respect; he must earn it.

(b) To provide for a sense of belonging

A narrow interpretation of a sense of belonging can be achieved if the manager is 'nice' to his subordinates, treats them politely, listens to their complaints and advises them on their personal problems. A broader and more meaningful approach is based on two main aspects:

1. information, and
2. recognition of endeavour.

A manager should provide sufficient information for each subordinate to have a proper understanding of why the objective he has to meet is necessary and how it fits in with the work of others in the enterprise. This will engender interest in the problem and allow the subordinate to use his initiative in co-ordinating his work with others.

Everyone likes their good work to be recognized : this does not need to be done publicly, but recognition should be given only after a proper assessment of the work done against expected standards. The standards of performance should be high ; praise for reaching a standard which the subordinate knows is low will not be very highly valued by him. In assessing a man's work a manager must also be direct in saying that the work done has not come up to expectations, but blame and criticism should never be given in public.

Apart from encouragement in his day-to-day activities, a subordinate's sense of belonging to the organization is enhanced by an annual or biannual formal appraisal of his overall performance and an assessment of 'how well he is doing' in the organization. This should be a personal assessment. If declared publicly, it may disrupt any team activity and destroy the subordinate's loyalty to the manager, e.g. 'he thinks more of Smith than me'.

A common fault of some in management positions for the first time, either due to lack of thought or from a misguided desire to further their own interests, is to take the results of good work by a subordinate to their own superior and not give credit to the man who did the work. This lack of integrity will lose that manager any respect otherwise earned. How easy it is to say 'this is Jones's work' ; better still to have Jones in on the discussion.

If a man starts to develop a sense of belonging through the work he does, he will probably develop an interest in the wider activities of the enterprise. This interest can be greatly extended if the manager can, through informal conversation advise the subordinate of the problems, hopes and aspirations of the enterprise (or his part of it), and in particular give early warning of any changes in the organization which might affect him. Some managers fear to do this ; 'these matters are confidential' ; 'no business of those I direct' ; 'why worry staff with things that may never happen'? There is an element of truth in all this, but in my view nothing encourages the sense of belonging more than the giving of this type of informal information. In giving his trust, the manager can earn respect.

(c) *To provide for achievement*

The first essential in providing for achievement or self-fulfilment is the delegation of responsibility. In delegating responsibility a manager must also delegate authority : responsibility without the authority to carry out the tasks involved is meaningless. However, in delegating responsibility and authority the manager is still accountable for the actions of the subordinate. Many managers are too cautious of delegation ; they complain that their supervisors do not delegate enough and yet are overworked because they consider their own subordinates incapable of assuming responsibility. To push responsibility down the line is the best way of tapping the initiative of everyone in the organization and of keeping interest in their jobs. The manager must strike a balance between retaining too much himself and overloading subordinates beyond their present ability.

Other managers claim that people fear responsibility, but DRUCKER (1955) feels 'it is not a matter whether the worker wants responsibility or not. The enterprise must demand it of him. The enterprise needs performance : and now that it can no longer use fear it can get it only by encouraging, by inducing, if need be by pushing the worker into assuming responsibility'. A real 'high-flyer' will take responsibility before it is given and on his promotion everyone says 'and about time too'.

The second and more important aspect in creating an environment for achievement is participation—in these days rather a hackneyed word but a keystone to modern management. Participation gives the individual the opportunity to respond to the challenge of freedom in using his initiative and creative abilities, and leads to job satisfaction and high morale.

The management process is based on a systematic methodology : analysis, decision, implementation and appraisal. Since the earliest days of scientific management the analysis of the problem, planning of what to do and how to do it has tended to be divorced from the implementation. It is now recognized that motivation of peak performance is achieved if subordinates are consulted in the fixing of objectives and have responsible participation in the planning of the work they do. In effect the manager delegates the planning of the work, but asks to be consulted : given that he has delegated the responsibility for planning (and implementation) he has also given his subordinate the authority to decide what to do. The manager must resist the temptation to overrule a good decision by a subordinate with an only marginally better one by himself based on a personal preference.

However, responsible participation may not be applicable at all levels of management ; some subordinate groups may be confused by it and be looking for guidance and decision by the manager. It is part of the manager's competence to adopt a flexible approach and to decide the balance needed. The essential requirement is that subordinates see the end result of their efforts and thereby get a feeling of satisfaction in their work. Another common fault in management is the confusion of 'participation' with 'consultation'. Many managers, particularly in deciding policy matters, quite properly seek the views of, i.e. consult with, their subordinates ; this is not participation since the subordinate in this case has no responsibility for his views (indeed management cannot delegate its responsibility for defining the overall objectives and policies of the enterprise). Also, the 'selling' of a decision already taken by management would be classified as giving of information in order to achieve a sense of belonging ; it is not participation.

MANAGEMENT ATTITUDE

Fundamental to the management process is the personal relationship between the manager and his immediate subordinates. The quality of the process greatly depends on the attitude of the manager in his day-to-day dealings with his subordinates. Given that an authoritarian approach is generally no longer practicable, successful management must depend on replacing the traditional worker/boss relationship, which was based on each side maximizing its own individual gain, with a relationship which sees the success of the whole enterprise as its objective. The essence of such a relationship is that a manager should trust his subordinates; if he does not he will find it difficult to delegate responsible participation and so get the best out of all those engaged in the enterprise.

An attitude of trust inevitably leads to a less formal relationship: subordinates are encouraged to regard themselves as colleagues in the endeavour, to work with the manager rather than for him. Some older managers may feel that with this approach they would be striving for the popularity of their subordinates rather than their respect. Respect for a manager by a subordinate is an essential part of the relationship, but it is earned (or not) regardless of how formal the relationship may be. To earn respect a manager must first and foremost know his job and have some knowledge of the specialist skills of those he is directing. He must also be a competent manager; far too often those who achieve a breakthrough in the field of human relations lose the benefits through lack of organization, system and control. Lastly he must be scrupulously fair, consistent and genuine in all his dealings with his subordinates. A manager must remember that every person is different, each with his own feelings, and must respect each subordinate as an individual. By the influence of his own conduct and good example a manager will contribute much to the training and development of his subordinates as future managers.

It is appropriate here to consider some of the do's and don't's to be followed in trying to foster better personal relationships in management. In the giving of information, do not talk down to your subordinate; when he consults you or you seek his views, listen rather than do all the talking yourself. In discussion, act the devil's advocate, question his wrong ideas and encourage him to think and work it out for himself, rather than imposing your own judgement (this is an important part of a manager's responsibility to develop the management qualities of his subordinates). A manager has failed if, in each situation a subordinate says 'now what does my manager want me to do here' rather than 'what do I think ought to be done'. In the giving of directives, avoid edicts and orders ; the same results can be achieved by requests without them being regarded as a weakness in your approach. In all dealings be courteous and considerate, and maintain a calm approach no matter what the stress of the moment. In criticising the work of a subordinate, a 'bawling out' has no place in modern management. There is no need to raise your voice to assert your position ; an erring subordinate should be encouraged to think out what went wrong for himself.

In considering the standard of behaviour expected of the modern manager, it is clear that integrity of character is the essential ingredient ; known personal dishonesty or suspicions of malpractice will form an insuperable barrier to the development of morale. It is now recognized that the ability to lead, i.e. to motivate others to a peak performance, is not a right acquired by accident of birth and education, nor is it a technique that can be wholly taught. It does require certain natural talents, intelligence and quality of character, but, given these basic assets and a willingness to recognize the human aspects of the management process, a man can be taught management techniques and can develop into a good leader.

The picture drawn is one of a manager as a paragon of virtue ; the objectives are difficult to achieve in any aspect of life, but worth striving for if any manager is to get the most personal enjoyment, satisfaction and reward out of the enterprise he is managing.

ADDITIONAL RESPONSIBILITIES OF TOP MANAGEMENT

'Top managers' are those who decide, amongst other things, the overall objectives and policies of the enterprise. Such persons have additional responsibilities, since their attitude permeates throughout the whole organization and confidence in and respect for them will largely determine the morale of those involved in the enterprise. These responsibilities also apply to levels of middle management which, with respect to those directed by their subordinates, are in 'top management' positions.

Organization in the management process defines the formal contact between manager and subordinate. (Although many organizations transgress the rule, it is generally recognized that there is a limit to the number of subordinates who should report to one manager and that desirably this limit should be about six.) The organization will also provide the framework for delegation of the various responsibilities to various levels of management. It is important that instructions, consultations, etc., should be given via the lines of responsibility and recognize the relationship set out in the organization structure. A manager should not bypass a subordinate and deal directly with those further down the line without having the subordinate present or at least informing him immediately he has done so. This is a common fault of top management, which quite naturally feels it can give instructions to anyone, often without regard for the priority of the work.

'Management is a continuous activity and cannot be replaced by techniques and systems designed for operating in the prolonged absence of the manager' (BRECK, 1967). In this respect there is a responsibility for top management to have some contact with, and in particular recognize the efforts of, as many of those down the line as possible. We have all admired the ability of certain top managers to maintain personal contact with those in their organization who are not directly responsible to them and whom they see but infrequently. They have an unerring memory of people's names and always some recollection of the personal problems or interests of each individual. This may reflect a particular aptitude, but it may also be the result of an effort to be so informed. There is some doubt that persons in top management positions make sufficient time to acquire this ability.

In some organizations, regular but informal meetings are held between top management and those working in all, or a section of, the enterprise. The main purpose of these meetings is to try and create a sense of belonging by the giving of information about the organization. Whatever the means, the object of a top manager maintaining contact is for him to be recognized and known within the organization as a person, even with good and bad qualities, and not just a 'god on high'.

ON BEING MANAGED

All employed persons will find themselves at some time in their careers being managed by incompetent managers. Much can be gained by such persons trying to analyse what is disliked in the management process, and to learn not to make the same mistakes when they are in a management position themselves. In order to do this it is desirable to have some knowledge of management and consider some of the above attitudes and responsibilities in reverse. It is as important for a subordinate to have trust in his manager as *vice versa*. If a subordinate does not trust the management to give him a fair slice of the cake and to provide him with continued opportunities for advancement, then he should leave the organization.

The importance of being informed has been stressed; a subordinate should see that he has sufficient information for him to have an understanding of the management's objectives. If the information necessary is not forthcoming, he should ask for it. It has been said that 'no one looks after you quite so well as you do'. This is true but in seeking better reward, information or opportunities from management, a subordinate should adopt a balanced approach. He should not bottle up his worries so long that they explode out of him. Do not hold a gun at management: 'do this or I leave'—the second alternative will more than often be taken. Give managers a chance to improve their methods. Many have learnt more about management process through talks with their subordinates than from guidance from the top.

At some time in their careers, those being managed may have to assess the value of forming and joining staff groups, associations or unions, one of whose functions is to communicate with and obtain information from management on behalf of the group. In large organizations, local government, etc., such associations exist and have an important part to play, particularly in the negotiation of salary structures. I suggest that in a modern management process, which recognizes the importance of good communication, the need for such associations may be less. The individual may find he loses his identity and the benefits gained by group action may be geared towards those of average performance and militate against those with real ability.

The giving of information is also a two-way process; subordinates who have been delegated a specific responsibility should keep their managers informed of the more important decisions taken and actions carried out. Again, a balance has to be struck. Many subordinates feel they can succeed in the organization only if they continually place themselves before the manager; this often means that the manager still takes the decisions and the experience of responsibility is lost. Others feel that in being given responsibility they should get on with it and not consult the manager at all. If in doubt the subordinate should go to his manager and say 'this is the problem and this is how I intend to solve it'. A subordinate should recognize when a top manager has bypassed his immediate superior and advise his superior of the situation. Also, if the right level of mutual trust and understanding has been built up between manager and subordinate, it should be possible for the subordinate to discuss with the manager even such matters as the opportunities for alternative employment outside the organization. Such discussions at least allow the manager a chance to give the subordinate some information on the opportunities within the organization, so that the subordinate has sufficient information to make a proper choice.

A manager who provides his subordinates with the opportunity to acquire a sense of achievement in their work also gives maximum opportunity for the individual to satisfy his ambitions. Having achieved one target in his career, it is right and proper for anyone to set himself another and strive towards it. This process can, however, lead to personal stress and an unhappy life. Each individual should strike a balance in the assessment of his own ability, setting himself the challenge of reasonable objectives whilst avoiding sitting back in any easy job and resisting change.

Given that no managers are perfect and that it is almost impossible for a subordinate to remove a manager, subordinates who are generally satisfied with the enterprise should adopt a tolerant attitude towards the worst foibles of the manager whilst the slow process of education proceeds. Those in a subordinate capacity have almost as much responsibility for encouraging good management as the managers themselves. Although in this paper aspects of the manager-subordinate relationship have been emphasized, the subordinate-subordinate relationship within a team is equally important. In all his dealings a manager should avoid taking actions which impair relationships among individuals in his team, and subordinates should recognize that the success of the group activity is in their interests. Differences of opinion between colleagues must be resolved with tact and understanding and must not lead to 'feeling' or 'friction'.

IN CONCLUSION

In the past more progress has been made in the civil engineering industry through improved equipment and technology than with the better use of human resources available. The contribution of good managements has only recently been explicitly recognized. The growth and prosperity of many organizations is sometimes threatened by the shortage of good managers. Traditionally managers have obtained their expertise by experience—by learning from their mistakes. We now require management responsibilities to be assumed at a younger age, and it is appropriate that efforts should be being made to include the subject in the formal education of the engineer. However, all the education and management skills will be of no avail unless the manager has a personal integrity on which to build his relationships with his subordinates. Likewise, even managers with the highest integrity and natural ability to deal with people will not succeed unless they have the skills to establish the necessary organization and management systems to run their enterprise. It is therefore right and proper that this conference should be considering both aspects.

REFERENCES AND FURTHER READING

BRECK, E. F. L. (1967) *Management: its Nature and Significance*. Pitman, London.

DALE, E. and MICHELON, L. C. (1966) *Modern Management Methods*. World Publishing Co.

DRUCKER, P. F. (1955) *The Practice of Management*. Heinemann, London.

HERZBERG, F. (1959) *The Motivation to Work*. Wiley, New York.

3

COMMUNICATION AND PERSONNEL STRUCTURE

D. J. Coats and J. Woodward

To conceive, plan, design and construct civil engineering works in a time of rapid change such as the present, requires not only clear thinking but swift development and execution. To allow this, each party concerned must be aware of the function and objectives of the others, and the relationship between parties must be not only clearly defined but also readily understood. Such definitions are clearly stated in *Civil Engineering Procedure* published by the Institution of Civil Engineers in 1963 and we will not therefore report them here. Also, although we are aware of other relationship patterns we will confine our discussion of communication and personnel structure to the promoter–engineer–contractor relationship culminating in construction under the ICE *Conditions of Contract*. For this, the promoter or employer must agree with the consulting engineer on what form the project will take and what programme is required, and the contractor must be given full opportunity of pricing the contract realistically and executing the works efficiently. Since, in practice, the three parties concerned are not individuals but organizations, their structures should be compatible and the channels of communication clear. Because the practice and problems of each of the parties concerned will be touched upon in other papers we intend concentrating on matters concerning the interrelationships.

PERSONNEL STRUCTURE

The objective of communication is primarily to control, and control implies authority and responsibility, which in turn require structures or patterns of personnel. In this sense, therefore, good communication depends on a good personnel structure and such structure should be so designed as to provide good communication so that every important task is done, but done only once. While a good structure does not guarantee good communication it makes it more possible.

Organizational systems and charts can be very fascinating, but it must never be forgotten that they concern people. It should also be remembered that management is as much an art as it is a science with no universally accepted body of authoritative principles, since the complexities of human nature are not as amenable to discipline as are the physical properties of inanimate objects. However, we suggest that the following matters are too often lost sight of:

(a) Personnel structures are not devices for creating status, but are only one of the means of assisting the efficiency of the firm concerned. There seems to be evidence that adoption of the principle of least number of levels of management is beneficial. This recommendation is called by other names such as the 'short chain of command' or the 'flat organization', but essentially it means that there is little point in creating positions or levels in management beyond the minimum. Peter Drucker has said that 'the growth of levels is a serious problem for any enterprise no matter how organized. For levels are like tree rings; they grow by themselves with age. It is an assiduous process, and one that cannot be completely prevented.' This may be true but it should not be accepted without a fight. Certainly, the scattering of meaningless titles should be avoided, if for no other reason than to ensure that a promotion is truly significant. However, the main reason is to allow direction without too many intermediaries and so avoid dilution or corruption of instructions or information between initiator and final recipient. Because a civil engineering project has a limited site-construction time there is usually little opportunity for the levels of site management to 'grow like tree rings'. The resulting organizational simplicity is of inestimable benefit to the industry, and many of the organizational pitfalls on which more sophisticated industries founder are avoided. We therefore suggest that simplicity of site organization be preserved, although modern techniques of management as well as of construction can still be utilized to the full.

(b) At every level of management responsibility and authority must be matched. There is little point in a man being responsible for something without the authority to ensure that it is done. Similarly there is no point in having a man with the authority to do something without his being finally responsible for the result. Job descriptions should therefore make the responsibilities and authorities of the position quite clear and an internal 'flat' organizational structure helps in this. Clarity is also required in interorganizational relationships.

Under the ICE *Conditions of Contract* it is accepted that there should be no direct contact between the contractor and the promoter or employer, although the contract is between these two parties. The engineer is given the responsibility of controlling the project on behalf of the employer, although he also has the duty to keep in close contact with the employer, both to ensure that the employer's requirements are being incorporated in the project and to keep the employer fully informed on the contract. In this connexion, there is great advantage in the employer making one of his own staff particularly responsible for liaison with the engineer during the development and construction periods. It is not necessary that he be an engineer although if he is this is helpful, but he must be sufficiently senior to be able to make decisions which will be upheld by his superiors. We are not here talking about a project manager employed solely as a coordinator for a particular project since such a man would neither have the responsibility, the authority nor the knowledge of the employer's requirements to act effectively. We are referring more to such as new-works engineers of public authorities or development engineers of private concerns, who are clear as to how the particular project fits into the whole scheme of requirements of the employer and who have a detailed understanding as to the use to which the project will be put by the employer on completion.

As to the engineer-contractor relationship 'there should exist a strong bond of common interest between the engineer and the contractor, because both should wish to see good construction materialize and both want a successful issue to crown their labours' (as *Civil Engineering Procedure* puts it). Certainly, the concept that the engineer and the contractor are on opposite sides of a prickly fence does not encourage the smooth operation of a project and is never to either's advantage. The ICE *Conditions of Contract* clearly sets out the duties and responsibilities of both and recognizes an engineer's representative (or resident engineer) and a contractor's site agent as the responsible people on the construction site.

During construction the engineer's representative and the contractor's agent, if not the only channels of communication between these two parties, must be fully informed as to all such communications, and it is usual practice for the engineer when writing to the contractor's head office to send copies of his letters both to his representative and to the contractor's agent. Similarly, communications between a section engineer on the resident engineer's staff and his opposite number are usually routed through the resident engineer or a copy

C

of any letter is given to him. When the resident engineer and the agent are both responsible chartered engineers of experience, they should be of great assistance to one another. Both should have the progress of the project as their primary concern and any action which is not towards this end is to be regretted.

(c) Although personnel structure indicates the main channels for the flow of information flexibility of operation must always be retained. Informal internal coordination or horizontal communication should always be allowed if not encouraged, and on *Figure* 1, which illustrates the structure of one consulting engineer's office, these relationships are shown by broken lines. By such means special knowledge is not locked up in one division, but it must always be remembered that while these informal consultations are of great benefit they have no bearing on responsibilities which are shown by firm lines.

Similarly, in the site relationship between a resident engineer's staff and the contractor's organization, while the only authoritative communication is between the resident engineer and the agent, informal assistance at lower levels is essential and important matters are confirmed formally later.

(d) Any personnel structure should allow the creation of a climate in which people obtain satisfaction and can develop their own abilities. Such a climate does not usually flourish in a rigid structure where the needs are moulded to the structure rather than the structure to the needs. Change must always be possible and continuous consultation between levels of management is necessary to ensure that changes are made when the need arises.

The unique nature of civil engineering projects where every job is different allows flexibility of personnel structures from job to job. Also the physical presence of a project itself and its particular needs provide the obvious objective which will determine the personnel pattern.

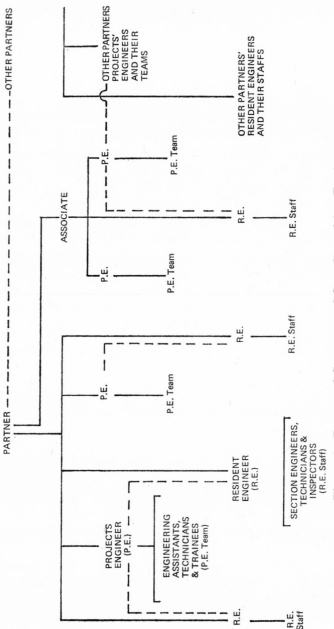

Figure 1. Personnel Structure of One Consulting Engineer.

The organization required for a motorway project will be different from that required for construction of a water-supply filter station, and each presents opportunities for engineers and others to be given a measure of responsibility different from that previously carried.

It has been shown that job satisfaction and motivation are more dependent on job content than on context factors such as supervision, relationships with others, work conditions or even salary. In civil engineering we are fortunate in being able to give engineers in particular not only interesting and varied work, but also a considerable measure of responsibility at a relatively young age and, because of this, little attention has been given to the mechanics of motivation in the industry. However, with all sides of the industry becoming grouped into larger units this subject may have to be given more detailed attention soon, for if motivation is not followed by the opportunity for satisfaction frustration will follow.

HUMAN RESOURCES

The proper use of human resources, the optimum balance between technical and non-technical staff and the development of technical staff in the best interests of the industry require study. Some of these aspects will be referred to by others, but comment is probably pertinent here.

The function of non-technical staff is well enough understood. The role of the professional engineer is being more clearly defined as the range of specialized knowledge expands, and the concept that, wherever possible, non-technical duties should be carried out by non-technical staff so that technical staff can concentrate on technical matters is increasingly being accepted. However, between highly qualified chartered or potentially chartered engineers and non-technical staff there lies a gap at present inadequately filled by the small number of civil engineering technicians. This is an unsatisfactory state of affairs which requires more energetic attention than it is receiving at present.

COMMUNICATION

As has already been said the object of communication is to ensure that every important task is done but done only once, and this implies control. For proper control, of course, one must have real understanding of the underlying intentions and therefore a second objective of communication is the passing on of information. Also, the recording of fact is essential to control, and such feedback is therefore important. *Figure* 2 shows the main communication pattern for a construction project. Communication should be specific, concise, unambiguous and

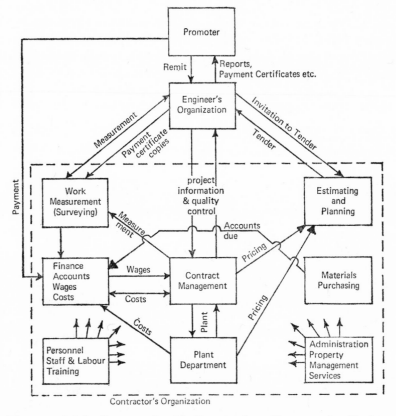

Figure 2. Main Communication Pattern for a Construction Project.

relevant. This can be effected by careful selection and use of alternative media for particular purposes and such media can be verbal, written or visual.

Verbal communication, although convenient and often helpful in developing a sense of team action, can be treacherous. Personal discussion or telephone conversations allow exchange of views and save time, but conclusions should always be confirmed in writing. Letters, reports and other written communications should be logically compiled and well set out using short sentences where possible, since they will be read by busy men. Most junior engineers would make better letter writers if they avoided circumlocution and increased their vocabularies. By using forms for routine work the benefits of written and visual communication can be combined, but the art of preparing meaningful forms should not be underestimated.

Visual media of communication range from bar-chart programmes to reinforced-concrete detail drawings. They are probably the most readily appreciated by engineers. Recent developments such as network analysis have received much publicity, and standardization of drawing terminology, symbols and techniques is being encouraged. However, a drawing, for example, must never be considered as an end in itself. It should be prepared with the recipient always in mind. A drawing accompanying a report to a client will have to be intelligible to laymen and reinforced-concrete details might have to be read by steel-fixers working in the wind and rain.

PROJECT DEVELOPMENT

A consulting engineer will have detailed consultation with a client often extending over a period of years to ensure that the design will match the employer's requirements, and such consultation will continue during the period of translation of design into construction. These consultations are best conducted at the level of a partner of a consulting firm and a senior executive of the client with only detail discussed at a more junior level, and in order to maintain this channel properly most consulting engineers insist that all correspondence from their offices is signed by a partner. But the client is not the only party to be satisfied during the development of a project. Approval has to be obtained from interested or affected parties such as government departments, planning authorities etc. It is the consulting engineer's duty to provide his client with such drawings, estimates and other engineering documents as are necessary to obtain these approvals at the appropriate times, and, if required, to act on his client's behalf in submitting proposals.

THE BASIS OF A CONTRACT

An invitation to interested contractors to tender for a contract should firstly convey an adequate description of the nature, scope and extent of the works to be executed, the method of construction where appropriate, the standards of materials and workmanship to be achieved, and the conditions likely to be encountered. Secondly, the enquiry should be in such a form that the documents can be used as a firm basis on which the contractor can quote a price for the contract, and so that competitive bids from a number of contractors may be quickly and effectively compared. Thirdly, the enquiry documents should be suitable for use as the basis of the contract so that they become contract documents.

To meet these requirements civil engineering tender documents usually comprise a form of tender, conditions of contract, a general specification describing the works and conditions at site, a specification of materials and workmanship, a bill of quantities and a set of tender drawings. The form and content of each of these parts is the responsibility of the engineer acting on behalf of the promoter, but it is generally helpful if a standard or familiar pattern is adopted. To this end the Institution of Civil Engineers and others have prepared a *General Conditions of Contract* and a *Standard Method of Measurement* and the Ministry of Transport, for example, has published a *Specification for Road and Bridge Works*. Due to continuously developing technology and the peculiar needs of particular projects specifications have to be especially compiled for most contracts, and it is most important that they are adequate since the contractor has little or no opportunity of amending inadequate documents at the tender stage. If this initial step in communication between engineer and contractor is faulty only trouble can result. The contractor is required to enter unit prices in the bill of quantities and to extend these to produce a tender amount. If these prices are to be properly prepared the specification drawings and each bill item description and quantity must be realistic even although the work is subject to re-measurement on completion. A contract document usually makes provision for negotiating rates or prices during a contract where variations have arisen, but since these are usually contentious matters they are best avoided, if possible, by initial accuracy.

COMMUNICATION DURING CONSTRUCTION

The predominant means of communication from engineer to contractor during construction is the issue of working drawings showing required construction details and appropriate written statements on work sequences and other requirements not specifically covered by the contract documents. The assessment of progress and discussion of points of difficulty are the subjects of regular site progress meetings attended by senior representatives of the engineer and the contractor. Such meetings should be controlled by an efficient chairman and a record of decisions taken and views expressed should subsequently be sent to all who attend. An important aid to the efficiency of site progress meetings is the existence of a detailed programme which is clearly marked or coloured to show progress to date. As has been mentioned earlier the engineer will keep the promoter informed on progress.

During construction the contractor is naturally anxious that he receives payment for work done both in fair measure and promptly. Under the terms of the contract the contractor usually submits his valuation monthly and the engineer advises the promoter by issue of a payment certificate of the sum which he considers proper at that time. To avoid discrepancies between the valuations of contractor and engineer detailed measurement of works completed is often done jointly by site staffs, but pending completion of this process estimations are often acceptable to prevent delay in interim payments.

RECORDS

Even such a brief review of communication in civil engineering projects would not be complete without some reference to those records normally prepared at site. The contractor will wish to use the experience on one contract to his advantage on another, and to allow this his site staff will be feeding information to his other departments as *Figure* 2 illustrates. Such information will include records of costs, methods of construction, plant suitability and site circumstances. The engineer will want to provide the promoter with 'as-built' or record drawings incorporating notes and details which will be helpful for operation, maintenance or extension of the works. Both the engineer and the contractor will want to establish the correct valuation, in terms of the contract, of the works completed.

To ensure that all this is possible information of a historical, quantitative and qualitative character must be recorded and, where possible, agreed by all parties concerned at the time. Unfortunately site engineers often regard this duty as a chore which only interferes with the interesting process of 'getting on with the job', but this attitude is all too often regretted when final settlement of the contract is being discussed or when the users of the works require information affecting operation.

INFLUENCES FOR CHANGE

As we have implied, the personnel structure and pattern of communication common on most civil engineering projects develop from the construction process itself and are not a little influenced by tradition. It is proper that all that is best in present arrangements be preserved, and basic and well-tried principles should not be readily abandoned. However, there are circumstances and influences at present appearing which may demand reconsideration of the position, and we would like to raise some of these briefly as the basis for discussion.

1. The advance of technology in civil engineering has already, in large measure, forced the demise of the small general-purpose consulting engineer and, on the contracting side, the number of specialist subcontractors has multiplied. This process may continue and will have a marked effect on the organizational structures of firms concerned. Also, the increasing use of highly qualified technical staff by contractors is creating for these firms reservoirs of highly technical expertise which will demand an increasing dialogue between engineer and contractor, certainly during the construction period and possibly even before it.

2. As computers and suitable programs become more readily available the decision-making locus may become more centralized resulting in reduced flexibility at site level. Improved communication media such as direct-line television contact, teleprinter facilities, etc. may also tend to strip site personnel of responsibilities which may have a damaging effect on job satisfaction and motivation.

3. Regionalization and amalgamation of both public and private bodies tends to create situations where the promoter feels able to employ his own development and, possibly, construction staff. On the development side certainly there is an increasing tendency on the part of promoters to demand more detailed information during the development stage. The large part now played by government agencies in financing projects also means more widespread discussion of proposals at the development stage, and while these features are usually helpful in crystallizing the requirements of those concerned with a project they often lead to delay at the development stage. This greater employer-participation also often means a

desire in that quarter to dominate control during the construction period which tends to disrupt the traditional team-working of contractor and engineer.

4. It is most noticeable that the time span for projects from conception to completion is decreasing. Once a decision to proceed in principle has been made a promoter is impatient for work to start on site, and this puts a heavy burden on the consulting engineer if he is going to produce effective documents for competitive tendering. The contractor's answer to this is the 'package deal' which may have an increasing place in construction arrangements, but which can never replace the more satisfactory development and execution controlled by an independent professional engineer. However, perhaps there will be an increase in the number of contracts let on basic rates only, so that a contractor can be appointed at as early a stage as possible.

The time allowed for construction in particular is much shorter, encouraged both by promoter and contractor. The promoter has to consider the effect of interest during construction which at the present high interest levels amounts to about 1% of the cost of the works for every two months of construction time. In addition, he is anxious to use the project to produce income and any delay adversely affects this. The contractor, on his part, is greatly concerned about the effects of inflation on his costs and therefore as short a construction period as possible is to his advantage. Changes in personnel structure and communication methods may have to be introduced to meet this challenge.

5. While the rapidly developing technology tends to encourage changes during the course of construction, due to the introduction of new materials or the better appreciation of other factors, the shortening time span already mentioned does not lend itself to the incorporation of changes. The effect on a contractor's costs if orders or

negotiations on price with a subcontractor or supplier are affected by a change must not be underestimated. A better understanding of the problems of parties to a contract by each of those parties is necessary if immediate advantage is to be taken of new knowledge without deterioration of relationships.

6. Contractors are now having to carry large resources in costly plant and highly skilled technical staff, and, for proper use of these, continuity of work is essential. To allow this it is possible that national planning will increasingly determine the phasing of construction work and this too will affect the control of development work. Also, weather and other external factors must not be allowed to have their present disruptive effect, and dialogue between the engineer and contractor is increasingly necessary to allow this.

7. Although contractor's plant is usually very versatile it is probably most efficient on one particular operation. This means that method of construction must become a greater influence on design, as the cost and sophistication of plant increases. As a simple example, it is obvious that the size and weight of precast members must accord to the capacity of the handling plant readily available.

8. To ensure continuity of work, contractors are having to work over a wider range of locations which requires their staff and labour to be more mobile. This does not always suit men with domestic commitments, and this together with the unfortunate image (or lack of it) of civil engineering as presented to school leavers does not help to improve the quality of people entering the industry. This matter will no doubt be discussed in other papers, but it has its effect on the quality of communication possible and on personnel structures which are increasingly dependent on high-quality staff.

4

STAFF SELECTION

D. Shennan

As an introduction to the subject, it may be instructive to consider why we who are responsible for recruiting personnel have to do so. In other words why is staff selection necessary? There can be a number of reasons :

1. Your staff are leaving because of better opportunities elsewhere. This can mean either an improvement financially or in future prospects.

2. They may not enjoy the company of their fellow-employees, or the place of work or the type of work, or they may not like the way in which they are treated by their immediate superior.

3. You may in your organization have only a limited number of positions at any one level and an employee wanting to hasten his advancement may decide that the only satisfactory solution is to move elsewhere.

4. The person leaving may be unsatisfactory and you may be glad to see the last of him.

5. You form part of an expanding organization and extra staff are an essential part of this expansion.

THE NEED FOR FILLING POSITIONS

The mere fact that someone is leaving the organization does not necessarily mean that a replacement has to be found. The late employee may have been one of a team and may have been taking more than his fair share of credit, due to his colleagues or immediate boss covering up for him through a misguided sense of loyalty to him rather than to their firm. Alternatively, with complete disregard to all the latest techniques that have been carried out in management and control of work, it may well be that no one has appreciated that a smaller staff can now be adequate. Such situations must be subject to constant review in any expanding organization adopting modern methods if it is to be efficient. So it is important to assess the situation periodically and realistically before embarking on recruitment.

RECRUITMENT

Before you as an employing authority can select someone to fill a vacancy on your staff it is essential to have a number of interested applicants from whom you can make your choice. People of the right calibre are not necessarily looking for a new job at the time when you want them, and the ones who are on the market are possibly not the type you want.

As a prospective employer you are ready to offer what you can hope is an adequate reward for the most suitable talent you can find. Remember, however, that the applicant has his own problems : he is comparing the impression of your company gained from his interview with you with that of a number of others he is visiting. Somehow you have to sell your company in the same way that he has to sell his services. He must feel that there are prospects ahead of him, and that such ability as he has will be encouraged and developed so that he can hope to advance progressively in his career. Somehow, therefore, you have to persuade the right person to apply to you. This involves making your staff needs known. The most usual way is by advertising either in papers which have nationwide or local coverage, or in selected technical journals. Advertising is expensive ; so it is most important to ensure that your insertion is properly planned and set out, and interesting. There is a lot of 'know-how' in preparing a bold and effective advertisement, which catches the eye. It is instructive to look through the staff vacancy jobs of certain well-known papers. If you compare these you will find some attract your attention and some you hardly notice. To give some general idea of cost, a comparatively small notice say 2 inches square might cost anything between £4 in a provincial paper and nearly £50 per insertion in one of the national papers. For more senior positions a larger display is required, which will cost proportionately more.

It must be borne in mind that some time must elapse before any advertisement can appear. This could be, under normal conditions, in the region of a week, but in a monthly journal might exceed 6 weeks. It is also important to have a sound knowledge of the effectiveness of the various media, since an advertisement in the wrong paper or journal can produce negligible results.

An alternative source of recruitment is from one of a number of agencies. These generally charge a percentage, which can range between 5 and 7½% of the first year's salary, so the cost is roughly similar. A major source of recruitment is from the universities and colleges. This can be initiated in a number of ways:

1. Probably the simplest and most effective is to publicize your firm's openings and prospects in one or more of the careers productions which are issued periodically.

2. A simple, but likewise very effective, method is to ensure that hand-outs are available to those who may be interested.

3. Visits by one or more senior members of the recruiting organization provide a better opportunity for both employer and employee to decide about each other.

Recruitment need not be confined to graduates, particularly nowadays when so many sandwich students are looking for industrial training. There is a very good case for engaging sandwich students in their first period of training, and subject to their receiving good reports from their place of work, re-engaging them in subsequent years to complete their industrial

g, and eventually after they have graduated. This is of
:age to the student, in that for some years ahead his
~~~~~yment in progressively more demanding work is assured,
and to the employer in that when a student returns he already
knows something of the firm's organization and staff, and so is
much more useful than a newcomer.

In the case of postgraduate recruitment these men are
required for the more senior posts and will have obtained
practical experience since graduation with some other company.
Generally speaking this type of recruitment is confined to
sudden surges in demand for staff as when one or more large
orders are awarded almost simultaneously. It is, of course, more
satisfactory to encourage staff in your own organization, where
over a period of years they have become well established. They
know the firm's method of working and they are loyal to the
firm's interest. On the other hand, there is much to be said for
having an inflow of new ideas and varied experience from other
directions, since this helps to avoid a possible tendency to get
into a rut.

INTERVIEWING AND THE INTERVIEWER

Assuming that your recruitment campaign has been satisfactory, you will be inundated with letters and telephone calls
from interested parties. A process of 'weeding-out' is then necessary. A certain number can be rejected out of hand as being
quite unsuitable. From the remainder a selection must be made,
and to these standard application forms should be sent for
completion and return. A further selection will then be possible
for actual interview.

Quite a lot has been said and written about interviewing. The
first and all-important objective must be to set the interviewee
at ease, so that he can talk freely and do full justice to himself.
Particularly for a young man, what is probably his first introduction to the big world of business can be an alarming
experience. It may be essential therefore for the interviewer to
do most of the initial talking, and apparently irrelevant chatter

on his part may be all-important. As soon as possible, of course, the talking must switch to the interviewee, by finding a subject in which he is interested sufficiently to forget himself and start expressing his views and recounting some of his past history.

Motivation should not be overlooked. Why does the applicant want to leave his present firm, why does he want to join yours, and why did he initially adopt his present career?

It is essential I think, that the man who is to be a successful interviewer should like people and be interested in them and have a sound knowledge of human nature. In the case of technical staff recruitment he should preferably be a senior member of the firm who has himself done the job for which he is interviewing. The basic principles can be taught, but experience is probably the main quality required. It is essential that any interviewer carries out a systematic appraisal of his past assessments so as to measure the correctness or otherwise of these.

STAFF SELECTION

My own concern is primarily with technical personnel of all grades for the management of the larger type of civil engineering contract, and the remainder of these notes will be biased in that direction. The man in control of such a construction site is known as the agent. The staff selected for the junior grades will in due course provide the recruiting ground for the agents of the future, so it is relevant to consider the qualities required. Of the many necessary the three most important are probably drive, decision, and human understanding.

The agent on a large civil engineering contract has to be the leader and manager of some hundreds or even thousands of men and may control the operation of several £100,000's worth of plant. He needs drive to ensure that work is carried out expeditiously and economically, because a firm of contractors cannot stay in business unless it makes a profit. The ability to make well-judged decisions is an essential quality in any leader. This does not mean being dogmatic, but when a crisis arises—as it inevitably will—he must have the ability to collect opinions

D

and information and advice from the various sources available to him. Then he, the man in charge, must decide which of the alternatives is to be adopted. This positive and decisive facet of leadership is one to which the men working under him will respond.

Without human understanding no one can hope to get the best response, enthusiasm, and loyalty from his staff. The old idea of two separate divisions of management and workers is out of date. The agent, the leader, must form a team which includes every single person on the site. This means that everyone—including the man with the pick and shovel, and the tea boy—can be made to feel that he is a valued member of the team. Proper development of this basic approach in human relations will produce a happy, contented job where people enjoy their work and carry it out with enthusiasm. With a job running under these conditions (unless the estimate has been a very poor one!) it is difficult to avoid making a profit.

In interviewing students and graduates, therefore, one is looking for these potential qualities, drive, decision, and human understanding. The first-class honours man may have his head too much in the clouds. The ordinary-degree man, and particularly an extrovert type who has taken part in team games and possibly been a prefect at school, is more likely to have the ability to accept people and to be acceptable to them. It is here that the civil engineering technician has an opportunity to excel. Such a man, with a less academic background but possible greater practical gifts, can well overtake his more theoretically trained colleagues. There is, thus, great scope for such young men to spread their wings and develop, at an early stage, real responsibility.

TRAINING

In some branches of industry the graduate can almost immediately be placed in a position of responsibility, because he has already carried out very similar work at college. In the contracting world, however, this is seldom the case, and the graduate—or for that matter the student—has to adapt himself to a situation where, if he is honest, he is forced to admit he knows remarkably little. His degree is initially of little use because of the wide gulf between the rather rough and tough contract existence and the comparative serenity of academic life. During his first year or two on a contract the young graduate has, therefore, to be given a comprehensive training to develop his practical gifts and teach him how to cooperate with other people.

The three main aspects of site work in which he can hope to gain experience are (1) technical (2) administration, and (3) human relations. Some training can, and probably is, given in these subjects at universities, but the effectiveness is many times greater if carried out under contract conditions, where, for instance, information is required by night to enable work to continue next morning. The sense of urgency, the feeling of being one of a team, and the satisfaction of jointly producing something cannot be taught in any classroom. The overall effect of contact with men of many types with a major common interest produces a broadening of character which is of lasting value to the individual.

I have dwelt on this aspect of training because in selecting one's staff one has to try to engage men who will respond to such opportunities. Such experience is, of course, essentially aimed at providing a sound grounding in the basic aspects of site work, which is required for the Institution of Civil Engineers Professional Interview.

LEVELS OF RESPONSIBILITY

As a young man amasses knowledge and experience so he becomes ready for greater responsibility. An important aspect of staff selection is to have a regular system of appraisal of existing staff's capabilities, so that when vacancies occur reference to the records may well show someone ready for promotion who can be considered, instead of recruiting from outside.

In some spheres it can be difficult to assess relative levels of responsibility. In the case of contract technical staff, however, there is an automatic progression of responsibility for all men on site who have the necessary qualities. Starting as a trainee engineer, as he gains experience he is graded as an assistant engineer and then on passing his Professional Interview he is called engineer. The further steps in progress are through section engineer to senior engineer, sub-agent, agent, senior agent, and area manager.

SALARY

Adequate payment is of course of first importance and it is essential that the offer made and the prospects of future increases bear a reasonable relationship to prevailing market conditions. However, money is by no means everything, and if a keen young man knows that he will receive considerate treatment with ample opportunity for development of his latent talent, he may well consider that this is more than adequate compensation for a slightly lower salary. This will be particularly the case if there is the prospect of stimulating and interesting work, and agreeable associates and working conditions. The danger in this last case, of course, is that having obtained his sound basic training and becoming a useful member of your staff, he will then realize that he has a market value in excess of what you are paying him, and he will leave to join a more generously minded firm and you have to start recruiting again.

As responsibility increases so, of course, must salary, and it is therefore essential to have a realistic median salary scale. This is to a large extent related to age, because this, at least in part, should correspond to increased experience. In addition this must be augmented for more than average responsibility, and correspondingly reduced for the steady plodding type who may still provide a very useful service in the right position. In establishing this scale information has to be collected from various sources, and one must try to adopt a median which is not too high to use up a large proportion of your profits and not too low to drive your best men away to your more generous competitors. But it must be emphasized again that the retention of a keen enthusiastic and loyal staff requires more than adequate remuneration.

ADVANCEMENT

In any progressive concern there should always be an opening for a young man of over average ability, which may be either theoretical or practical. If you do not promote him he will become restless and will probably seek promotion in some other company. If you wish to retain him let him see that you value his services, listen to his opinions, and let him feel he is part of the organization. Select him carefully, pay him adequately, treat him properly, and he will be one of the mainstays of the organization in the future. The importance of understanding, appreciative, and humane management cannot be over-emphasized.

In most large organizations there is a family-tree structure with a large number of people in the most junior grade and numbers steadily reducing as responsibility increases, until at the top there is probably one man, the chief engineer, or managing director. If, therefore, one starts with say 100 trainee engineers, at each stage as these men progess (and one hopes becomes more competent) there are progressively fewer vacancies. Some stay for years in junior positions, some leave for other companies, and a comparatively small proportion only can be promoted. One big reason why so many young men are lost to the company of their first choice is that periodic appraisal of their potential has not been carried out. It has, therefore, been left to some other firm to appreciate their worth and induce them to leave by the promise of a higher salary. It need hardly be emphasized that each time a staff member leaves and has to be replaced this costs money, and can involve duplication, wastage, and delays.

APPRAISAL

There are a number of ways of approaching this subject. The six simplest, and probably within limits the best, is to have a series of say six qualities such as industry, intelligence, potential for advancement, human understanding, reliability, and have five grades each ranging from very good, through good and average, to below average and poor. It will be found that the majority of reports tend to take the middle range, but whereas one may be above average, another relating to the same man may be below. It is, therefore, important to know the persons assessing so that their returns can be weighted as necessary against those from other sources, so producing a fair and balanced over all report.

CONCLUSION

Finally, I would like to emphasize again that when you are selecting staff they in turn are selecting their future company through you. Just as they are trying to sell themselves to you, so you have to make an effort to sell your company to them.

5

TRAINING FOR MANAGEMENT

G. S. Bosworth

IN ANY discussion about management one is bedevilled by the problems of definition. It is a term which is frequently used as a synonym for business ; in fact on occasion I have heard under-graduate studies referred to as business studies and post-graduate work on the same topic referred to as management studies. From the point of view of the man on the shop floor the management is a term descriptive of the grey, faceless ones who stand between him and his true reward ; to the young executive the term represents status or a position in the hierarchy ; it is the equivalent of commissioned rank in the Forces. In post-war years it has come to be identified by quite a number of people as a profession ; indeed a professional institution exists on this basis.

I consider all these definitions to be wrong and misleading because I am quite convinced that each one of us, consciously or unconsciously is engaged in the practice of management just as we are concerned with breathing, or happiness. It is some-thing we all do, and it runs through the whole pattern and activity of our lives. I would like, therefore, to analyse this concept briefly and draw your attention to some of the implica-tions which arise from such an analysis.

The successful practice of management requires competence in three areas. Firstly, the ability to identify objectives in the environment in which management is being practised, and following on this the ability to set those objectives in order of importance. Secondly, it requires the ability to marshal and deploy the available resources to the optimum achievement of these objectives. Thirdly, it requires the ability to measure the extent and efficiency of achievement of the objectives. Put even more succinctly the practice of management is concerned with the optimum use of available resources to achieve desired ends. This is almost identical, at least in intent, with the statement in the constitution of the Institution of Civil Engineers which states its purpose as being—'To direct the great sources of power in nature for the use and convenience of man'.

Of course such near epigrammatic statements give an air of simplicity to what is a very complicated subject, and I would like now to look a little more closely at the three areas which I have defined. The ability to identify objectives in the environment must be heavily dependent upon knowledge and understanding of that environment. As far as the individual is concerned, the environment starts with him, the place where he lives and works, the community that he works in, the organization which employs him and extends outwards to the world and beyond. There are no real sharp divisions in this, it is a spectrum-like continuum and beyond a certain point moving outwards we all live in the same environment. This is another way of saying that we are all citizens whatever our particular profession or interest may be. Whatever we as individuals do in our own particular corner must ultimately have repercussions on the total environment, and vice-versa. I suspect that this is a fact which is often forgotten or ignored; were it not so, the worst effects of the industrial revolution on our environment would not have been allowed to occur. Consequently, the engineer is not wise to confine his search for objectives merely to his immediate surroundings, or the purely technical problems of his profession.

Of course, knowledge and understanding can only be acquired if the individual is perceptive. The objectives are undoubtedly in existence, but if they cannot be perceived, then their presence cannot be noted. Consequently, to knowledge and understanding must be added power of perception, and there are those who might argue that these two are in some measure incompatible. In other words too much attention to study and learning may dull what might otherwise be a lively perception.

Assuming that we have identified some objectives, then comes the problem of setting them in order of importance. This calls for those intangible qualities of judgement, wisdom, experience, and is, I suppose, one of the areas where this new management technique of decision-making is required. Clearly, there are quite a number of objectives where the factors may be quantified, in which case decision-making can be accomplished by computers, but I suspect that very many are surrounded, if not penetrated, by imponderables, where one is concerned with quality rather than quantity decisions, where instincts and emotions as well as reason must be brought into play. Ability in this field seems to me to be a compound of experience in the environment, knowledge and understanding, intuition and instinct. It is one of the principal areas where the art of management is all-important.

Having selected the objectives, that is to say having decided what to do, then it is necessary to set about achieving them, and I suspect that the first step in most cases is to create a climate which is favourable to the achievement of the objective. In other words all those people who might bring influences to bear on the operation should preferably be disposed to work for its success; at least they should not be opposed to it. This is the area of public and human relations and it is often a crucial factor in the success of an enterprise. Apart from this, however, it is necessary to marshal and deploy the available resources. In very general terms they are usually classified as manpower, money, equipment, materials, and so on. They differ in kind and quantity from one environment to another. Both plumbers and surgeons use tools, but they are different. Manpower in a research laboratory is different from manpower in a foundry; men to a foreman are Tom, Dick and Harry, but to a chairman they are a productivity resource measured in terms of cost and output. It should not therefore be surprising that a man who is successful as a foreman may not be successful as a chairman of a company or vice-versa. The resources available to each of them are different, and are used differently. To use any resource effectively and efficiently one not only needs to be knowledgeable about it, but also to have skill and imagination in its use, which in most cases can only be derived from constant practice, preferably under the guidance of a skilled practitioner. One has only to think of the difference between knowing the rules of cricket and being able to play cricket well. Undoubtedly, there is a requirement of innate talent, but even the most talented will profit from good training and even the least talented will be better for having practised under a skilled instructor.

To almost every young man the real symbol of the practice of management is the control of other men, usually in a hierarchical sense, mainly concerned with directing their manual or routine mental activities. This is a relatively simple concept and gives rise to standard drills and disciplines for the accomplishment of defined assignments ; but today, if not always in the past, the important manpower asset lies in the brains, ingenuity, innovative power and ideas. This is a most powerful resource and one of the most difficult to manage since it does not easily respond to the sort of disciplinary control which used to be necessary in the armed Forces. This problem is not simply one of human psychology, that is to say, making people want to do the sorts of things which need to be done. It is a much more difficult organizational task of how to create a corporate man, that is to say, how to get ideas processed by a group of people. These are the problems which probably stand out quite sharply in the management of a research department or a design office, and have given rise to new techniques of brainstorming or thinktanks as experiments in corporate thinking. Similarly there are problems of logistics, not only of material things, but also of processes which have to be managed and they have given rise to techniques of PERT and its variants.

Within a given environment, therefore, it is necessary to know and understand the resources which are available and to have practice in their deployment in the achievement of objectives.

The third area of competence is that of being able to measure the extent and efficiency of achievement. In some cases this is relatively easily done. One can usually see that a job has been completed, though not always. In financial terms it is often possible to measure the efficiency of achievement, but there are many factors where ordinary physical measurement is not practicable, such as happiness, or morale, inventiveness, or even productivity. Sometimes in these cases it is possible to say that an existing state of affairs is better than a preceding one without specifying precisely by what amount or on what scale. Lest you should think that this is slightly fanciful you might care to consider on what criteria you decide that one man is better than another, or that one man/machine-tool unit is better than another; or how you measure the efficiency of a design office. Difficult though some of these comparisons may be, it is well to avoid selecting apparently measurable criteria such as the possession of academic qualifications, or perhaps even more doubtful, years of experience, as units. This is not to decry many of the techniques which are in existence, such as merit rating, work study and work measurement, job analysis, and so on, but this is an area where, in my view, much important work remains to be done, and there is not a great deal of time in which to do it. I suspect that using concepts like productivity, without clearly identifying who may improve the situation and how the results are to be measured, may lead us to more confusion than such incentives were designed to clear up.

I have taken a great deal of time in trying to define what is meant by the practice of management and the sort of things which are necessary for it to be practised successfully. If you accept my definition then training for management should start as early as possible and go right through schooldays onwards into adult life; a major part of the training is concerned with ensuring adequate knowledge and understanding of the environment in which the individual is living and working. In a general sense this is one of the prime objectives of all education, but each one of us lives in a specialized environment and it is necessary to consider how far the formal educational system equips us to deal with that. The young person leaving university may be expected to have a fair knowledge and understanding of the general environment, and quite a lot of the engineering science which he will need as a background to his professional occupation. He will, however, need to become acquainted with the organization within which he works and the people who work alongside him. He will have to overcome some inhibitions which may have been implanted accidentally by the educational system; he will, for example have been taught by means of examples drawn from an idealized world of weightless rods and frictionless pulleys. He will probably have been encouraged to solve most problems by the creation of mathematical models and he will have acquired the habit of solving problems on his own because of the examination system. Thus he will be predisposed to be a solitary operator. Moreover, the problems that he has encountered usually had finite right answers, whereas problems in real life are not even confined to the particular topics which he will have studied, and there will be no right answer, only an optimum within the many constraints imposed by factors which previously he has not had to consider, such as time, money, human reactions and so on.

In the main he will only have acquired significant knowledge and understanding of the natural phenomena and natural resources considered to be important in his particular branch of engineering. It is very unlikely that he will have had any experience in organizing teams of men and I doubt very much whether he will have had any personal experience of working as a member of a team solving a complex general problem. It is virtually certain that he will have had no experience of creating what is in effect a corporate man or organization, which is designed to carry out on the large scale those activities which on a small scale are carried out by one man.

How then is one to establish a training programme in the practice of management? The first step is clearly to identify the knowledge, understanding, skills and techniques which are necessary in a particular appointment. Starting from the beginning this is derived from the definition of the objectives of the enterprise as a whole, because these determine the organizational structure or division of tasks which are necessary to accomplish those objectives. The traditional analysis into posts on the organization chart, such as chief engineer, sales manager, research director, and so on, are not really adequate for this purpose; they need to be expanded into a task analysis in order to ensure that, on the one hand, all the functions required to be discharged by the organization as a whole have been accounted for, and, on the other, that the responsibilites of the individual are clearly defined. This can then be set against what the candidate for a post could normally be expected to have and the difference represents what needs to be accomplished in the training programme.

This is an irksome process, because it involves looking at the whole organization, and it is quite understandable that many training schemes consist of a modified form of apprenticeship, which in most cases means putting the young person in a number of departments and hoping that he will pick up what he needs to know. This is rather like sending a potential fireman to look at a lot of fires in the hope that he will pick up the necessary drill. It must not be forgotten, however, that during a tour of this kind, however prolonged it might be, the young person will pick up a great deal of information about his environment and he will also learn a lot of the tricks which the firm would rather see eliminated than perpetuated.

I have to confess that, never having worked in the field of civil engineering, I could not tell you how I would set about devising a training scheme for management in this field of activity, but in the electrical and mechanical manufacturing industry where I worked for some thirty years, it was clear that it was not possible to allow young people to practise on a real-life situation in the early stages. After giving some thought to this we were able to devise, some fifteen years ago, a very reasonable facsimile of a real-life situation under the title of 'Design-and-make projects'. In these schemes we provided a group of young graduates with the facilities of a drawing office and workshop, and a customer in the form of a works manager, or research director, who required some piece of equipment for which he was prepared to allocate part of his budget and which must be in his hands within three months, or thereabouts. He was thus in a position to specify, like any industrial customer, what he wanted and what he was prepared to pay for it. Possible solutions to his problems would be produced by the team and, like any sales organization, they would have to tell him their preferred solution, if they could. They then had the problem of detail design and manufacturing within a cost and time limit, and finally had to satisfy the customer of the satisfactory performance of the equipment. By this means we created the total ideas-to-hardware situation within which all these young people would ultimately work.

This is the first introduction to the practice of management, and becomes a grid of reference for the future. After some experience on the job in various departments we considered that it was desirable that the young person should be introduced to the elementary concepts of business management, so that he might understand the overall constraints within which a business necessarily operates, though he himself will be probably concerned with only one department in that business at any one time for several years to come. Usually, coupled with this business training there would be instruction in the management problems within his chosen department or function, and we found that this was difficult to provide in the form of courses in the educational system since very few seemed to exist. However, I am sure that the need is there, and that the supply will grow to meet it. From this point forward only those who are going to be concerned with the business management of the total enterprise need further exposure to business management courses such as those in the major business schools, but of course, from time to time all those concerned with management need to be brought up to date on the techniques available for data-handling, measurement and communications.

Perhaps the most important thing to remember in all discussions on management education and training is that, whilst the principles of management practice are universal in their application, the environmental factors and the resources available differ from one field of operation to another, and different resources need different techniques in their handling. It is for this reason that I do not believe that there is a profession of management, though there can be a profession of business management, or a profession of estate management, or works management. For these reasons the liberal application of courses of business studies will not necessarily improve the management of a particular business, nor will the learning of a catalogue of techniques and practices necessarily improve performance in a particular department, or function. Each programme has to be designed and built to suit the need, and perhaps the most difficult part of the whole process is to provide 'hands-on' practice in the management of a real situation. This requires not only skill in guiding the young person, but considerable courage on the part of the person doing the guiding and the people responsible for the overall operation of the enterprise.

6

MANAGEMENT OF THE DESIGN PROCESS

Peter Dunican

THIS PAPER concerns the design of buildings. This is an activity which involves a number of different sorts of technologies and technologists, which is an omnibus term to embrace architects, structural, constructional, mechanical, electrical and other environmental-services engineers, building economists, and quantity surveyors, etc., and what they do. The paper is written from the viewpoint of the structural engineer, although it is not intended to be partisan or parochial or prejudiced, difficult as this may be to avoid.

The design of buildings is a multi-disciplinary activity, which traditionally has been directed, rather than managed to any formal management plan, by the architect in a simple direct and separate relationship with each of his technologist advisory consultants. Building design has generally been approached on an individual basis, in the same way as each building has been dealt with, seemingly in isolation, as an individual project. Until recently it has not been accepted that the design of buildings is a process which is susceptible to the processes of formal management techniques. There were three main reasons for this, firstly we did not know exactly what design meant as an activity; secondly we did not know how it was carried out, despite the fact that management is a function of design, and design is a process of management; thirdly, the design of buildings is an art as well as a science dominated by the idiosyncracies of the designers. We know that structural design is particularly concerned with imposing order and method on construction and, like construction, it is a process which can be managed, although it is generally true that most designers do not recognize the managerial nature of the process which they are carrying out. This is also another primary difficulty, even if it is a psychological one.

Clearly, to manage we must know more about design as an activity or process, and how it is carried out and, perhaps what is more important, how it should be carried out to make it more efficient. If you do not know how you do whatever it is, and what is wrong with the way you do it, and its consequences, how can you possibly improve on what you do and, perhaps what is more important, on what you are doing it for? The main reasons for trying to improve the way we design is to reduce the effort and improve whatever it is we are designing. One definition of design is the solution of a problem within a given resource envelope; and management must be concerned with making the most efficient use of the available resources to produce what is required at the right time, at the right price, with the minimum effort. This requires all concerned to work closely together; that is, it requires team working.

Team working to be successful requires that the members of the team must not only be competent, but that they must also be compatible, be able to communicate with each other, and above all care for what they are doing. They must be totally committed to the task in hand. As design teams consist of people, so must management be concerned with how to get them to give of their best, and for each and every one of them to derive the maximum satisfaction—material, intellectual, professional and spiritual—from their work. The individuality of the team members is most important—the strength of a team can be more than the sum of the strength of the individual members of the team—but for success it is necessary to recognize individual strengths and weaknesses, because by doing so any unbalance within the composition of the team can be reduced to the minimum, that is if it cannot be eliminated.

E

Broadly, design teams can be set up in a variety of different ways, but in the design of buildings there are two main possibilities. Firstly, and the more usual, is through a series of separate individuals or groups, brought together under the leadership of one or another of the team, but usually the architect, to deal with a specific problem. Or secondly, and not so usual, is through a multi-disciplinary group or entity, which has within it all the necessary disciplines or skills required for dealing with the design problem. In each case the management problems are the same in principle, although they are different in detail, mainly because of the different communication problems.

Communication is an integral part of the management process, and it is essentially a two-way system, which must ensure that the required information of the necessary quality and quantity is transmitted and received by all concerned at the right time. The right information at the wrong time can be even more frustrating than the wrong information at the right time; so therefore the timing of communication is also important.

What studies there have been of design clearly indicate that it is a process of :

Analysis—defining the real problem, collecting all the relevant information, and listing and evaluating all of the conditions which govern the solution ;

Synthesis—producing all the solutions which satisfy the governing conditions ;

Evaluation—determining which solution meets the needs of the problem in the best way ; and

Communication of decision—to all concerned in the form required for the decision to be properly understood and acted upon.

This process is carried out in a series of defined stages, beginning with the strategical concept leading to the tactical detailed instructions required by those commissioned to bring about its ultimate realization. Ideally, this cyclic process of design would be an even, uniform, progressive continuum. In practice at best it moves forward unevenly and irregularly, and at the worst it moves forwards and backwards through the cyclic process seemingly in jumps and jerks.

Any plan for the management of the design process must clearly recognize the inherent creative function of design, which although susceptible to rationalization cannot be preempted. As Chermayeff has said, inspiration is a special moment in a rational process. Of course, to have any inspiration you must be capable of being inspired, but unfortunately this is a state which cannot be preordained or predetermined. It is probably true that in the past, the need for inspiration has been overstated in the design of buildings, but nevertheless inspiration is the hall-mark of good architecture. It distinguishes the good from the commonplace—and there are too many commonplace buildings being designed and built today.

A number of attempts have been made to set up management plans for the design of buildings, but broadly speaking they have not been successful because of their failure to appreciate what is good and what is bad about the present way of carrying out design. Nevertheless in 1967, following preliminary studies made by the Tavistock Institute and the Ministry of Public Building and Works, the Royal Institute of British Architects produced their Plan of Work, which essentially was a management plan, to deal with the problem in some detail. Its stated aim was to define a systematic method for the design team to deal with their particular problems. It separates the technical and managerial function of the design team, and it is concerned primarily for use by design teams made up of separate individuals or groups. It is not conceived to improve the standard of design, but this will undoubtedly be one of its most important fringe benefits, because by allowing the more effective use of the time and the resources available it should lead to the possibility of carrying out design more rigorously. The Plan is not perfect by any means, but it does establish that the design, as it is developed, passes through a series of defined stages, and that at each stage it is necessary to carry out the design decision sequence and formally confirm its completion before moving on to the next stage (see diagram 1). The plan cannot deal with all difficulties, but neither can we unless we identify them. From the structural engineer's viewpoint the major deficiency in the Plan is its lack of appreciation of the contribution which the structural engineer can make in determining the constructional design.

It must be implicit in any structural or architectural design that there is a clearly expressed and understood method of building. The Plan does not appear to appreciate this fact. Another defect in the Plan is the implication that the architect is the leader and organizer of the design team. This is a difficult position for the individual architect to sustain, if at the same time he is personally responsible for the architectural design. This is because his emotional involvement in the design is likely to inhibit him in his role as team leader, where he would be required to make design decisions based on the advice given by other members of the team which may be in conflict with his preconceptions. After all, the design of buildings is essentially a compromise between what is ideal and what is possible, and it is here, of course, that one can be critical of the engineers and the other consultants, who appear to be insufficiently aware of the fact that their particular contribution is only a part of the whole, and that it is the whole building which is the most important consideration. However, experience of using the Plan—and given a reasonable degree of feedback—should result in the second edition being a great improvement on the first.

The structural engineer's involvement can be considered under three headings. Firstly, doing what is necessary to decide what it is he is actually going to produce, that is a compatible structural system for the building. This we will call the design. Secondly, carrying out the calculations necessary to justify the chosen structural system and provide the necessary information for the preparation of the instructions for the drawing office This we will call the analysis. And thirdly, producing in the

DIAGRAM 1.

BRIEF OUTLINE OF RIBA PLAN OF WORK

Stage	Scope	Usual Terminology
A	Inception Prepare general outline of requirements and plan future action	Briefing
B	Feasibility Provide client with an appraisal and recommendation in order that he may determine the form in which the project is to proceed, ensuring that it is feasible functionally, technically and financially	
C	Outline Proposals Determine general approach to layout, design and construction in order to obtain the client's authoritative approval of outline proposals	Sketch Plans
D	Scheme Design Complete brief and decide on particular proposals, constructional method, outline specification and cost. Obtain all approvals	
	Brief should not be modified after this point.	
E	Detail Design Final decision on every matter related to design, specifications, construction and cost.	Working Drawings
	Any further change in location, size, shape or cost after this time will result in abortive work	
F	Production Information Prepare production information and make final detailed decisions to carry out the work	

DIAGRAM 1 *(contd.)*

BRIEF OUTLINE OF RIBA PLAN OF WORK *(contd.)*

Stage	Scope	Usual Terminology
G	Bills of Quantities Prepare and complete all information and arrangements for obtaining tenders	
H	Tender Action	Working Drawings
J K L M	Project Planning Operations on site Completion Feedback	Site Operations

drawing office the necessary detailed instructions, working drawings and schedules for the contractor. This we will call the project management. The first stage is the most difficult one to organize at the present state of development, because of the degree of interdependence among the various members of the team, and the effect that this has on collective decision-making. The second and third stages—analysis and project management —are or should be most methodical processes and most capable of being organized or managed through defined procedures.

All of these notions clearly affect the organizational infra-structure of the practice of the consulting structural engineer. No doubt there are numbers of different ways in which this could be set up to deal with the demands being made upon it. However, for success whatever system is established must be flexible and capable of being adapted to circumstances as they change, because the prime purpose of any organization is to meet as efficiently as possible the needs it exists to serve, which by definition are external to itself. It is a fact that the trend is towards larger practices, and therefore it is appropriate to consider a system of organization which seems to work in such a circumstance.

Following a conventional management pattern, the practice is divided into a series of line and service or support divisions. Each division has a clearly defined function (see diagram 2). In this paper we are concerned with the work of a structures division (see diagram 3), which consists of about 40 technologists, that is engineers, 30 technicians and about 20 support staff. In the division the divisional management, that is the executive partner and the associates, are particularly concerned with :

Planning—what are the aims and *why*

Organization—who is involved and *how*

Directing—who decides *what* and *when*

Coordination—*who* keeps *whom* informed and about *what*

Control—who judges *results* and by what *standards*

Precisely how these management tools are used depends on the individual circumstances.

An incoming commission, or project, to the practice is allocated to a division, where it is placed in the complete charge of a project engineer, who is directly responsible to the project partner for all aspects of the project until it is completed ; the project partner is one of the divisional management. Directly the programme is known the project engineer will draw up in outline an internal plan of work with its manpower implications. This plan is modified as necessary following discussion with the divisional management, and when it is confirmed the necessary manpower is allocated, and a draft calculation plan is prepared. The calculation plan deals with all of the activities which follow from the design stage, that is when the structural system is determined in principle. In essence it covers stages 2 and 3 above.

In outline the calculation plan sets out what is to be done, the information which is necessary to do it, including the design conditions, the analytical techniques which are to be used and the sequencing of the production of the calculations and the drawings in the form of flow diagrams. The manpower

DIAGRAM 2

THE INFRASTRUCTURE OF THE OVE ARUP PARTNERSHIP

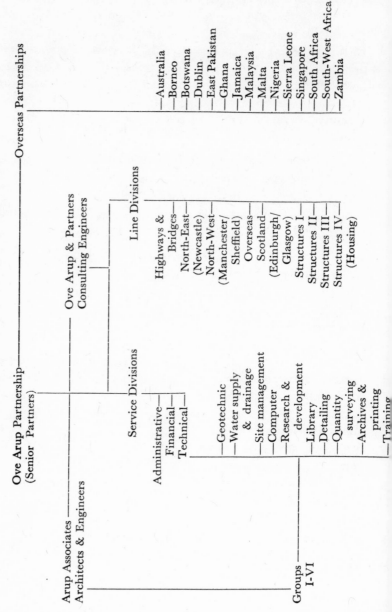

DIAGRAM 3

Structures Division: Ove Arup & Partners

OUTLINE OF MANAGEMENT FRAMEWORK

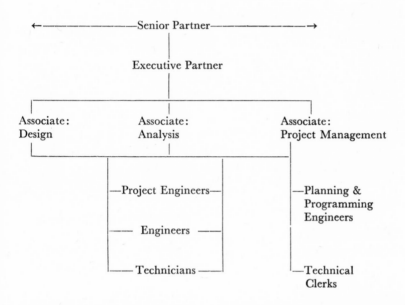

implications of the calculation plan are then finally determined so that they can be phased with the other commitments of the division. This requires a divisional plan. The divisional plan with its manpower implications has to be brought up to date as the work on each project in the division proceeds. Bringing up to date is difficult because progress on individual projects is dependent not only on the progress within the division, but also on the external supply of information which is not within the control of the division.

There are also the financial implications of what is being done. The division also exists to pay its way and to make a profit as well as carry out its technical commitments. This requires a responsive monitoring system which functions in parallel with the system of technical control. This is important because the quality of what is being done must not be inhibited by narrow financial implications. In other words the quality of the technical service must not be constrained by narrow internal financial considerations; this is the essence of professionalism and the true use of management techniques in the design process.

CONCLUSION

The crucial problem in the design of buildings is that it is divided amongst a number of different disciplines. But there is also another significant problem which has only been mentioned in passing, which is that design is divorced from production. In this sense building is rather particular and peculiar.

Generally, the way in which design decisions are made in a multi-discipline, generalized industrial situation is independent of what is being designed, because the producers are integral participators in the design process. In building, however, production is usually independent of, and separate from, design, and designers with perhaps the exception of the engineers involved are not knowledgeable about the processes of production. Therefore there is an inherent discontinuity—or credibility gap—in the building process if this is defined as all that goes on from when it is decided that a building is required until it is completed and ready for occupation. So, if you believe that design concerns directly the process of production, the situation in the building industry imposes an additional management problem of some significance. This is not an insuperable problem, but it is inhibiting, in that design decisions can more adversely affect the cost of the product, because if the design is not entirely right for production the cost must be unnecessarily inflated. And cost is money and money is part of the resource envelope. But the most important resource is man, and man by nature is not a collaborative animal, and generally the higher his intelligence the less collaborative he becomes. The difficulty is not primarily a technical one or concerned directly with money. It is social, emotional and psychological in origin.

It is difficult to prove the real advantages which are to be gained through proper management. So to begin with it needs an act of faith, and faith is not dependent on intelligence, but what we do know is that it is only through proper management that we can produce more with less, and therefore it is only by producing more with less effort that it will be possible for the needs of the people of this world to be met. This is particularly so in the construction industry and it is the ultimate justification for the management of anything.

LIST OF REFERENCES

BRUNTON, BADEN HELLARD & BOOBYER, Chartered Architects. *Management Applied to Architectural Practice*. George Godwin *for* The Builder Ltd., 1964.

CHAN, Dr. W. W. (1968) Programming techniques for the client, the designer and the builder. *The Structural Engineer*, **46,** 335–44.

CHAN, Dr. W. W. (1969) Programming techniques for the client, the designer and the builder. Discussion. *The Structural Engineer*, **47,** 181–6.

HIGGIN, G. and JESSOP, N. (1965) *Communications in the building industry: the report of a pilot study*. Tavistock Publications, London.

MINISTRY OF PUBLIC BUILDING AND WORKS. (1967) *R. & D. Building Management Handbook 3. Network Analysis in Construction Design*. H.M. Stationery Office, London.

MINISTRY OF PUBLIC BUILDING AND WORKS. (1969) *R. & D. Building Management Handbook 6. Programming and Progressing in the Design of Building*. H.M. Stationery Office, London.

REX, J. M. (1969) The management of multi-disciplinary design teams. Informal discussion. *Proc. Instn civ. Engrs*, **44,** 59–63.

ROYAL INSTITUTE OF BRITISH ARCHITECTS. (1967) *Plan of Work*. Reprinted from *Handbook of Architectural Practice and Management*. The Institute, London.

7

COST CONTROL IN THE DESIGN OFFICE

J. C. Judson

COST HAS always been of importance to the engineer, but traditionally his main effort has been concentrated on the design problems and the time-scale in which these can be overcome. This is still the case where new frontiers are being explored, as with the American space programme, but generally engineering technology has moved so far that the main challenge today is to complete the project within the time and cost allowed. Increasingly the engineer is involved in management, and the emphasis now given to the control of costs is producing marked changes of attitude in the design office.

The shortage of capital for new developments leads to a demand on the engineer to select the most economical solution and keep the cost of a job within the budget provided. Failure to achieve this can have a crippling effect on the owner who can rightly insist on the use by the engineer of an effective method of cost control.

THE DEVELOPMENT OF A PROJECT

In order to understand the role of the design office in the cost-control process it is useful to study the financial development of a project in the following stages :

1. Planning Stage—This is the phase in the development of a project which takes it forward from its initial inception as an idea up to the point where it is defined in sufficient detail for its viability to be demonstrated and a decision to be made to

allocate funds for more detailed development. The end-project of this stage is a feasibility report. The contribution of the design office will be to examine the technical solutions to alternative schemes and prepare estimates of capital cost and a programme for the phased expenditure of capital. The viability of the project may not be sensitive to the capital cost and the estimates will be very approximate but as realistic as possible.

2. Development Stage—This stage involves the design office in the preparation of an outline design from which a budget estimate and programme are prepared. On the basis of this the sources of finance are established and the decision to proceed is made. The budget estimate will not only indicate the anticipated total cost, but also give a broad indication of how the available money is to be allocated to the various sections of the project. It is essential to establish at the same time the extent of tolerance on the approved budget estimates that can be accepted by the developer. This budget together with a design brief form the reference point for the future work in the design office.

3. The Contract Stage—This stage takes the scheme through to the placing of contracts for the construction of the project. The design office must now produce documents for tender purposes. The accuracy of the tender design and the resulting contract price will depend on the programme requirements, which will define the amount of detailed design that can be carried out at this stage. The contract sum which may be made up from preliminary drawings, approximate quantities and provisional sums does not automatically become the project estimator's bible. However, the information gained in this stage, including the contractors' prices, will with suitable contingencies form the budget sum to be allocated to the project and the target for the cost control. The cost target must be divided into sections related to the programme as well as the developer's taxation, rating and other accounting procedures.

4. The Design Stage—The final stage to be handled by the design office is the preparation of the working drawings for the

complete construction of the project by the contractor. It is during this stage that the major cost-control effort is required by the design office. The design office must ensure that the financial consequences of the detailed design must not lead to costs in excess of that allowed by the budget.

5. The Construction Stage—This stage covers the fabrication off site and the construction on site to the completion of the project. More problems will occur in controlling costs at this stage, but they are dependent on the standard of the overall management of the project and the control of the site organization. Skill in the earlier estimating of contingencies to cover the inevitable disruptions that occur on site, and the careful control of expenditure against these contingencies are the key to this stage.

These stages may overlap and vary from project to project but the control of the capital cost follows a logical sequence throughout and leads to different problems in each stage. However, it is the early engineering and management decisions which often have the greatest effect on the project cost. Once the detailed brief and initial proposals are settled there are usually few opportunities for using alternative design solutions to effect savings, but many pitfalls that could lead to extra costs. If budget costs are being exceeded the only remedy may be for the developer to find more finance or redefine his requirements.

COST CONTROL

The progress of the project that has been described consists of a gradual refining of estimates until the eventual final cost of the project is known. The basic objective of cost control is to monitor performance against these estimates and to ensure that expenditure is kept within the money that is available for the project. It follows that at all stages the cost-control exercise must include accurate reporting against the estimates so that the necessary action can be taken. The cost-control system should also help to ensure that the developer's money is spent wisely and that the required balance of expenditure is maintained among the various sections of the project.

Any system of cost control must have the three basic principles of any control system which are :

1. a reference system ;
2. a method of checking actual performance ; and
3. a means of remedial action.

It will be noted that there are marked similarities with techniques used to control project time, and much of what is said about cost control applies equally to progress control. Time and cost are closely related on a project, and efficient management of the design office is to provide the project with the necessary control in both these areas.

ESTIMATES

Frequent mention has been made of estimates and they should be discussed at this point, since it is obvious that accurate estimating is the basis of successful cost control. This does not mean that detailed estimates are always necessary to ensure accuracy. Estimates are often produced from very little project information. As the accuracy of an estimate depends more on ensuring that it reflects the total content of a project rather than having the pricing exactly right, it follows that estimates will tend to show a lower cost than is actually incurred. This situation can be remedied by using historical cost data derived from past projects and converted into global costs that will include the numerous smaller items that are easily omitted when estimates are made solely from the detail available at the time.

Each stage in a project represents a complete concept lacking some detail and knowledge of unpredictable events which may occur in the future. As time passes the estimate is refined by taking note of progress in the detail design and events that have occurred on the project. Each estimate is therefore a blend of the costs of fully detailed sections of the projects, sums based on previous experience to cover the areas where detail design is not available, and a contingency to cover the areas where detail design is not available, and a contingency to cover the unpredictable element. Contingencies, therefore, play an important part in every estimate. They must not be abused and there must at the outset be a clear definition of what they cover and who controls their use. The developer, the estimator and the designer may each provide his contingency, and the danger of duplication must be avoided. The need for contingencies, however, must not be ignored. There are often strong political pressures to cut them, and the greatest danger to financial control of projects is ill-informed optimism in the early stages.

COST CODES

Successful cost control does not usually depend upon complex and sophisticated techniques, but upon the strict observance of relatively simple procedures. If they are followed from the very beginning until the end of the project the initiative is never lost, but if the routine tasks are neglected the source of management information dries up, and the project manager will have lost the initiative and henceforth be one stage behind events. To avoid this situation arising the first essential is to produce a project breakdown, which in broad terms will be used throughout each project stage. Every area and item can be coded. The detailed estimates, bills of quantities, contractors accounts, interim measurements, and final measurements and accounts then follow this project breakdown in order that all cost information can be directly related to the original budget estimate. Failure to arrange this procedure at an early stage will mean that costs will have to be laboriously extracted from each source and reallocated to the appropriate section of the budget before an up-to-date picture of the capital-cost situation is available. As with most laborious and routine tasks it is liable to be neglected and if this happens the manager will be deprived of essential information, so jeopardizing any control action he might need to take.

BILLS OF QUANTITIES

Bills of quantities form the natural basis for the target cost estimate and for operating the cost control. It should be stressed, however, that in their present form they have severe limitations, which if not appreciated can make them a dangerous tool. They are in need of improvement and the conventional bill is generally not completely satisfactory to any of its users, whether they be contractor, engineer or employer.

The usual project programme for industrial installations is such that accurate plant information is not available at the time the civil engineering contracts have to be placed. As a result the bills of quantities convey only an approximate idea of the nature and quantity of work that has to be carried out. They are then used to record the actual quantities of work, but are not sensitive to other important facts that may affect the contractor's programme and method of working, and consequently have an effect on the eventual cost of the project. The tender prices from such a bill do not therefore give a reliable guide to the final cost of the job, which, because of reassessment of rates and claims, may far exceed the tender sum.

As long as the problems are borne in mind at the time of preparation the experienced design office can successfully base the cost-control procedure on the bills of quantities. The bills should be as simple as possible with the minimum number of items, and then for even the most refined cost-control purposes it will be possible to omit many items that can have a small effect on the total cost. The bills should be produced as a logical extension of the earlier estimates, with an eye not only on convenience in measurement of quantities and settlement of a final account but for producing information for cost control. This means sections being related to the cost coding and to the stages and requirements of the project programme, as well as providing feedback for future estimating. To assist in identifying the effects of changes in programme the contractors should be encouraged to include in the preliminary items the various forms of fixed and time-related charges.

F

HOW TO OPERATE COST CONTROL IN THE DESIGN OFFICE

The most difficult case is on a large project with a tight programme so that the detailed design is carried out after the contract has been placed. Up to the completion of the tender a comparatively small team is employed. The cost control amounts to checking their current work against the previous designs and estimates, and reporting changes so that there is a full reconciliation between the first estimate and the target sum based on the tender drawings. The problem is to control the activities of the large group of designers, who are later engaged on the preparation of the working drawings. It is necessary that one controls costs and not merely reports the cost effect of the design, once it has been completed and there is no time for a change. This means that the designer himself must be involved in the cost-control exercise and conscious of the effects of his work. He will be primarily interested in the pure engineering and with little time he will be less interested in taking off quantities or preparing cost-control returns.

At the outset the designer must be given a clear design brief together with the tender drawings covering his section of the work. He must be trained to design within the concept of his brief and the tender drawings; then if he considers a change is necessary he must report to his senior. If a change is unavoidable a cost check must be made and the effect reported on the monthly cost-control sheet. This can be calculated as an extra or saving on the target sum.

As drawings are completed a check of the total cost is made so that as sections of the work are designed there can be a reconciliation with the target sum. As cost information becomes available in this way it is possible to assess how successfully the cost is being kept under control. Cost information is therefore available as early as possible, giving the manager the maximum flexibility in arranging remedial action which may in the extreme mean applying for extra finance.

Two types of cost check have been mentioned—one when the designer reports that he must deviate from the tender idea and the other on the completion of a set of drawings covering a section in the cost estimate. For both these the bills of quantities can be simplified to show, say, excavation, sheet piling, concrete and reinforcement for each section of the works. These quantities can be priced against global rates calculated to give the same total for the section as in the original bill. If a check against the simplified bills indicates a major change it may be necessary to refer back to the full bill. At the same time it must be remembered that quantities are not the only measure of cost, and the effect of changes on method of construction have to be taken into account. The check of quantities is within the competence of the designer, but to estimate the real cost effect of these changes requires special skills.

WHO DOES THE WORK?

The designer must take an interest in the cost-control process if it is not to degenerate into cost recording, but the preparation of estimates from engineering quantities should be done by a specialist. A solution is to have an estimating and cost-control section in the design office. They can be engineers and quantity surveyors with estimating experience, supported by juniors who can take off quantities. This section is responsible for the early estimates from the designers schemes, and for preparing the targets and cost control reports as a service to the project manager. During the design stage they assist the designers in assessing the effect of proposed changes in design and they are responsible for estimating the cost of the completed drawings.

The cost-control section must have the confidence of the designers and the further it is removed from the design office the less successful is the operation. The bills should ideally be prepared in the same office, but with a larger project it may be necessary to employ outside assistance, but this must be done in the closest collaboration with the design office, and the cost-control operation must take place in the design office. Nothing is more disastrous than the complete division between design and cost responsibility that sometimes results from the separate employment of quantity surveyors.

The responsibility for keeping the cost-control exercise moving rests with the specialist group, and this requires an active approach. If they wait for designers to fill in forms and return data they will obtain the wrong information too late. Only by chasing information, asking the right questions and probing in every corner will the estimator obtain a true picture from which he can prepare accurate estimates.

THE COST OF COST CONTROL

It is obvious that the organization and time proposed to exercise good cost control must add to the cost of the whole design process. One could expect to devote about 3% of the cost of design on operating a good cost-control procedure. The rewards to the efficient management of the project are obvious and in the same way that design organizations devote time to investigating alternative designs, so must money be found for cost control.

FUTURE DEVELOPMENTS

The limitations of bills of quantities have already been mentioned, and research is being carried out on possible alternatives to the conventional bill.* This aims to provide a bill that will give better feedback to designers, integration of bill data with planning and cost procedures, easier valuation of varied work, fewer claims and easier settlement of claims. The type of changes that can be expected are a reduction in the bill items used to describe the quantity-proportional changes. These changes would be grouped into activities where there are specific sequence limitations. In addition there would always be fully defined items entered by the tenderer to cover time-related and fixed charges. The standard method of measurement of civil

---

* CIRIA Research Project No. 98, University of Manchester Institute of Science and Technology.

engineering quantities is at present under review, and it is possible that this will move some way towards dealing with the problems arising from the present method of measurement.

As a next step one should examine the impact that the computer is likely to make on the design and management of engineering projects. In the purely design field programmes are possible that will optimize the design, using the bill rates, and produce quantities and the cost of the work. Where this is possible much of the work is taken out of the cost checking during the detail design stage. A further development is the 'total-systems' view where the total sum of information existing at any stage on a project is stored on magnetic tape. While this may not be possible with technical information one can visualize all information on interrelated matters such as the programme, the drawings, the quantities, the cost-control returns and design-office costs being recorded in this way. The effect of the completion of a drawing would then automatically be recorded against the programme, the final measurement and account, the cost control and the design-office costs budget, and given in the monthly returns on computer print-outs. There is much work to be done before the computer can lead to real savings on a large project, but progress is being made in certain of these areas. In the meantime there is still education required throughout the profession to make the engineer aware of, and interested in, his responsibility to control the cost of his designs and to appreciate the management problems involved.

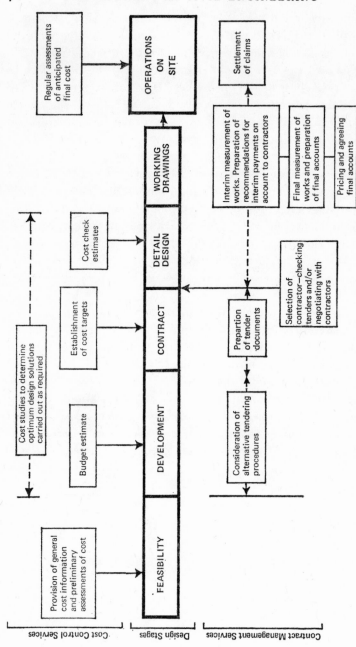

Relationship of Cost Control Services to Design Stages

8

MANAGEMENT AND COST CONTROL OF WORK ON SITE

J. Rowbotham

SUBLIMINALLY present on every site manager's desk is the thirteen-letter message, 'profitability'. That is why he is there, why his company exists; although we define management and spell out the processes of management readily, it cannot be overemphasized that site management, in whatever form it takes, must be established and operated to make sure that the project makes a profit. Since management of on-site resources accounts for by far the largest proportion of money spent each year on construction, it must be recognized that site management is one of the main problems of management in the construction industry. If proof were needed the sobering thought is that the bankruptcy list is usually headed by the industry and those who follow the company returns would hesitate to recommend the industry as the best of income investments. Construction can be a profitable and worthwhile business provided projects are managed effectively; this is particularly relevant in this competitive age for the small and medium contractors struggling for survival under the present economic pressures and against the big battalions.

Management as it is relies to a large extent on enlightened guesswork, experience gained over many years and the 'bar chart', and unfortunately the latter frequently is relegated to a corner of the site office and then forgotten. There are experienced and able site managers who can get things done and done well, without any but the most rudimentary aids. But, unfortunately, they are few and far between, and these soon end up in the boardroom or found their own firm. The gathering of experience and knowledge over a long span of years can no longer suffice as a basis for providing the skilled management in the industry—the pace and the degree of management experience required demands not only certain inherent qualities, but the expert knowledge and ability to make the best use and application of the many management skills and techniques that are available to the site manager. Management has got to become much more professional, in other words it is much more important to do things in the right way, use the best tools, techniques and methods. This is the basic need of the underdeveloped companies and is the central resource of the developed companies.

This paper is not concerned with the inherent qualities, the personal make-up or the technical competence required of site managers; this in no way implies that these are unimportant; on the contrary, these are the essentials that make the foundation of success in site management, and are dealt with elsewhere in this symposium. What this paper aims to do is to focus attention on the next major component of management competence —specific management knowledge—by reference to the demands placed on site management, and to the associated skills in certain functional areas of site management, and in particular, to the profit-preserving tool, cost control.

The key to profitability is productivity—maximum use of resources, manpower, machines, materials and money invested. Although the big management problem is that you are repeatedly plunged into the construction of a multi-million-pound contract, that has never been undertaken before, will never be built in that form again, and you never have the advantage of a familiarity curve, it is not a finite industry like canning baked beans; it depends on the ground, the weather and the client's requirements. Although these are variable there are management skills, techniques and operating procedures that can be applied to all projects, albeit with differing emphasis that will lead to better judgement and control in the process of site management.

SITE MANAGEMENT AND ITS RESPONSIBILITIES

For the purpose of this paper the functional term 'site management' has been chosen to indicate management with responsibility for the site and includes responsibility for construction management in respect of the work undertaken by the main contractor and carried out directly on site, control of subcontractors and specialists on site, general control of all activities on the site, and, according to the context, the term may refer to either (1) the field of site management activity, or (2) the line management personnel engaged therein. Such personnel, are usually site-based and responsible either to a principal (director or partner) or to a manager.

Only by examination of the demands—the range and degree of responsibilities of the man in charge of the site—is it possible to assess the knowledge and skill needed for the efficient performance of the management functions. These differ among firms, and are influenced by the size of the construction project, but in general it is possible to enumerate these responsibilities. On normal types of construction the site manager will be required to take full responsibility for :

Tender-stage activity.

Checking the accuracy and practicability of the engineer's drawings.

Negotiating with clients' professional advisers.

Detailed construction planning.

Master programme.

Site layout and organization.

Setting out.

Work study.

Safety.

Operation and maintenance of plant and transport.

Checking, recording and reporting periodically on (a) movement of materials and plant (*b*) operatives' time (*c*) sub-contractors work.

Site stores.

Payment of wages.

Type and number of salaried staff under his control.

Clerical staff recruitment, selection and dismissal.

Type and number of hourly paid supervisors.

Recruitment, selection and dismissal of operatives.

Operation of incentive schemes.

Negotiations with trade unions within the limits of the working rules.

Site training of students, graduates and other personnel under his control.

Control of quality.

Agreement of measurement of work done (including sub-contractors' work).

Certification of payments to sub-contractors.

Preparation and agreement of interim valuations.

Control of site costs.

Cost accounting on site.

Site progress meetings and minutes.

Preparation and agreement of the final account.

He will be required to contribute in respect of :

Liaison with the client.

Diagnosis of defects and agreement on remedial work.

Selection of sub-contractors and specialists.

Tender preparation.

Selection of method and plant and the design of temporary work.

Purchase of materials.

Hire of plant and transport.

Selection and dismissal of salaried staff for whom he is to be or is responsible.

In addition, he must have an adequate knowledge of the employer's technical and administrative policies and procedures.

KNOWLEDGE AND SKILL REQUIRED

Having stated the general range and degree of responsibilities to be undertaken by the man in charge on the site it is appropriate next to assess the knowledge and skill needed for the efficient performance of the management functions.

It appears fashionable today to restate some basic and sensible procedure or technique and give it a name or shroud it in clipped businesslike technique phraseology, and then to present it as the panacea for all the site disorders. Consequently there is a tendency for the industry to be overwhelmed by the variety of management techniques available and to be somewhat sceptical as to their merits, but projects are getting bigger and more complex, and something more than the haphazard methods of the past are needed if management is to get the required results from the available resources.

Management proceeds in a logical sequence :

1. forecasting, making plans, budgets and setting standards ;
2. organizing, motivating and coordinating the execution of the plan ; and
3. measuring and controlling performance ;

and we know that construction, no matter how small the project, cannot be effectively or economically carried out without these management activities—the four essential elements, planning, motivation, coordination and control.

In practice, these do not appear in so simple a form, but are the objectives of procedures, techniques and developed skills. Planning and control are the objectives of certain procedures and techniques, whilst motivation and coordination are activities specifically related to the personal skills. How these elements are matched in actual site management is illustrated by *Figure* 1. It will be seen from comparison with the schedule of responsibilities that :

1. The most important of the procedures and techniques of which site management must be aware are those dealing

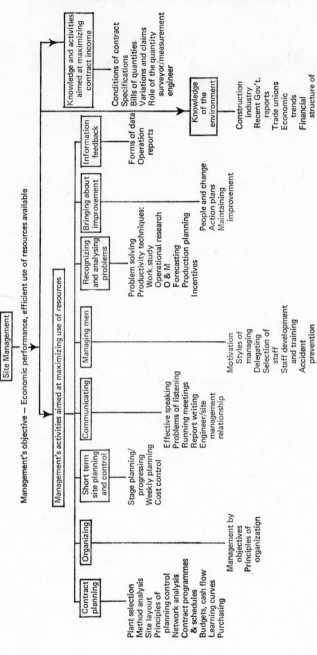

Figure 1. Site Management—functional areas and associated skills and techniques.

with planning, programming and progressing, cost control, productivity techniques, communication methods, and knowledge and activities aimed at maximizing contract income. These are the key techniques needed to direct any project execution.

2. The most important skills to develop are those relating to organization, knowledge of the environment, managing men and communications. Good knowledge of these being a prerequisite in successful site management.

Having established the procedures, techniques and skills required by competent site management I wish to refer in some detail to the need for, and operation of, efficient cost control, highlighting the attitudes and activities that lead to the escalation of costs and loss of profit, in an endeavour to emphasize the need—grafting business-school science onto engineering knowledge.

THE CONTROL FUNCTION

Most site management is conscious of the importance of planning and control—the necessity for a construction programme to be formulated and the establishment of progress control techniques to ensure the closest possible adherence to the programme. Unfortunately, site management has not the same depth of awareness of the importance of the planning and control of the financial resources of a project, nor is it familiar with routines that must be brought into being to ensure control of cost and spending. Control of performance and cost on the site, at the source, is vital. A successful contractor just does not happen ; it is the result of careful planning and rigorous control in accordance with a predetermined high level of efficiency. Site management, to implement the control, must have a knowledge of the financial structure of a construction company and know how the cycle of working capital operates. This in its simplest form is indicated in *Fig. 2*.

The amount of working capital required is determined by

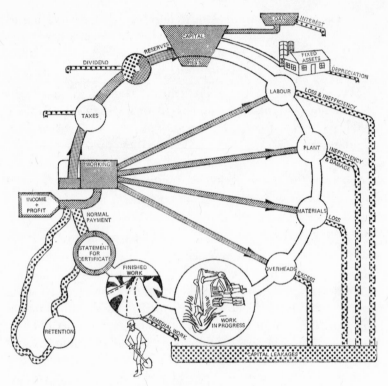

Figure 2.

Financial Structure of a Civil Engineering Company.

the absorption of the following element of total construction value at each conversion stage :

(a) labour,
(b) plant,
(c) materials,
(d) overheads, and
(e) profit.

From the system outlined in *Fig.* 2 it is seen that the profitability of a construction company depends upon the rate at which working capital is circulated—the more paid work that can be done with the same working capital in a given time, the greater the profitability. Site management must therefore minimize stocks of materials, work not eligible for payment, and the less outstanding debtors there are the more working capital there is for circulation. Fixed assets and costs (buildings, staff, overheads, etc.) must be watched for economical layout, utilization and efficient operation. In some cases the nature of the contract does not permit certain parts of the conversion process to be speeded up, but in general, the study of each of the various stages with a view to reducing the time taken to the very minimum will amply repay the effort.

Most construction companies rely heavily upon internally generated cash flow to meet debts as well as for financing growth. It is therefore essential that site management appreciates and is capable of establishing cash flow ; cash flow analysis enables site management to appreciate where to set up measurements to maintain adequate balance by controlling flow. This in turn allows profit forecasts to be made.

The flow of cash is a vitally important factor for all levels of site management. Too often there is an air of general secrecy associated with this aspect of a project, and the various levels of site management have little or no knowledge of how their particular contributions are affecting the process. The object must be to enlighten and impress that turning money to repeat the process must be done as quickly as possible for maximum profitability.

Profit calculated as a percentage of capital employed represents a broad guide to top management as to the overall efficiency of the company. Site management is concerned with profitability to the extent that it gives an indication of its own efficiency, and profit must be interpreted for site management in relation to the project execution. The illustration also shows that a site is like a colander where every uncontrolled activity is a hole for capital leakage. It is also seen that controls must be exercised by both site management and head office and for the latter to operate site management is required to feed information.

COST CONTROL

Whatever the size of the project may be, controlling profit and its constituents, income and costs, is a function of site management, but it is frequently considered that standard costing and budgetary control are useful and possible in the larger projects and beyond the resources of, and inapplicable to, the small project, since the establishment of a costing section is excessively costly, although such methods achieve considerable economies.

Even with the best of site management, projects of all sizes need efficient cost control to keep programmes on schedule and within budget, and a cost-monitoring system must go into effect immediately the contract has been let, and it must produce both weekly and monthly reports to site management on all critical areas until the project is complete. Without effective control costs escalate and it suddenly becomes too late ; site management must be warned before problems turn into disasters. The key to effective control is the establishment of a standard procedure and trained costing staff, preferably a cost engineer and certainly cost-conscious site management.

REPORTING THE PERFORMANCE

The principal instrument of control is represented by the weekly cost statement, comparing actual performance with the cost of the unit rates. For each operation, cost has to be clearly defined in terms of the elements that make up the cost:

1. Direct material costs—the quantity of material actually used in each unit of measurable work.
2. Direct labour costs—the quantity of labour used in producing each unit of measurable work.
3. Direct plant costs—the cost of all plant and equipment incurred as a result of producing each unit of measured work.
4. Indirect expenses—these are indirect costs, staff and other overheads.

The statements must be up to date and clearly expressed, and one of the most important responsibilities of site management, at all levels, is the analysis of variances between its unit rate and actual costs. The variances can generally be classified as follows:

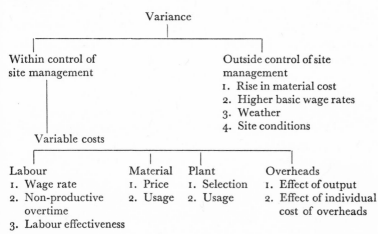

The variations have to be classified, so far as it is practicable to do so, as controllable and non-controllable, according to the responsibilities of the person in charge of the particular aspects; secondly a further analysis of the variances under the price and usage headings will be required. In theory it can be maintained

G

that all variances are controllable at some point in the organization.

The monthly cost statement consists of an analysis of all costs appertaining to the project, compared with a corresponding analysis of the interim revenue; the interim revenue is derived from the measurable quantities of work done, evaluated by the bill rates for each item. The precise form in which these statements are prepared, and the variances or cost headings which it contains vary according to the project and the needs of various site management to which the statement is addressed. The essentials are that the important variances from predetermined budget and rates should be clearly shown, and that the statement should be submitted promptly after the period to which it refers.

All management is based on comparison, and only by comparing what is happening at the moment with what happened in the past or with what was planned to be happening now can management make certain of the way the project is moving. Control statements such as outlined enable this principle—management by exception—to function by site management taking appropriate action to correct variations from the planned pattern. From the foregoing it must be accepted that the civil engineer in charge of construction-site management—has to be trained and to be qualified not only to design, but to manage in the business sense.

TRAINING AND DEVELOPMENT OF SITE MANAGEMENT

The success of a business, industry or any undertaking depends on the ability to develop people, and in particular its managers, and the success or failure of the construction industry is largely dependent upon the quality of its management. The National Board for Prices and Incomes in its Report on the pay and conditions in the industry stated 'The civil engineers who take charge of sites which may employ several hundred men need to supplement their professional training by careful and thorough instruction in many aspects of management including financial and budgetary control, and most important, the management of people'. It is encouraging to be able to report that this is recognized by the industry and particularly the contractor. The contractor, knowing that his only possibility of making a profit is in his management methods, and with the impetus of the Construction Industry Training Board, pays more attention to management training and development than any other sector of the industry, and it is he that will ultimately take charge—take over the role of total project management.

ACKNOWLEDGEMENTS

The author wishes to thank his college for permission to present this paper, and to express his grateful appreciation to the Urwick Management Centre and the Sundridge Park Management Centre for their permission to use to *Figures* 1 and 2 respectively. These are modified versions of charts prepared and used by the Centres.

REFERENCES

Cost Control in Engineering Management. Informal conference, University of Manchester Institute of Science and Technology, 1967.

INSTITUTE OF BUILDERS. *Construction Management in Building, Present and Future.*

NATIONAL BOARD FOR PRICES and INCOMES. *Pay and Conditions in the Civil Engineering Industry.* Report no. 91.

PARKER, E. J. (1969) The planning of project finance. *Proc. Instn civ. Engrs,* **43,** 261–71.

9

ORGANIZATION OF CAPITAL WORKS BY LOCAL AUTHORITIES

F. R. Dinnis

THE PURPOSE of this paper is to take a critical look at the organization of local-authority civil engineering capital works, and to outline some ideas for obtaining greater efficiency. The techniques of design and construction have made notable advances in recent years, but organization procedures have fallen behind and become complicated and cumbersome to a quite ridiculous degree due to many causes, the more important being discussed in this paper, which is divided into four sections : initiation, preparation, execution, and operation.

INITIATION

The first action to be taken following the feasibility study is the preparation of a preliminary report setting out the reasons why the scheme is necessary, with an indication of the likely capital cost, together with maintenance costs. Local authorities are notoriously prone to waste time on arguing about the need for a particular scheme. The report must therefore be positive, comprehensive, as short as possible, and must avoid alternative solutions unless these are obviously necessary.

Local authorities have to ensure that everything they do is within the framework of the law. Recent reports on local government have hinted at the desirability of a radical change in these matters, whereby local authorities are allowed to carry out any work which is clearly needed, even though not sanctioned by specific legislation. The present legal 'straitjacket' has a deadly effect on local-authority progress in provision of services.

Turning now to the political aspect, it is unfortunate that local authorities are mostly controlled on a political basis. It is frustrating to those of us in local government to find our schemes being delayed (and mutilated sometimes) by political actions. It is too much to hope for any change in the near future, but perhaps younger engineers can look forward to less interference from the new-style local authorities, which are now under consideration.

The part of the report which deals with estimates is most critical and difficult. An engineers, our instinctive action is to give realistic estimates of cost based on current or near-future conditions. However, almost without exception, preliminary estimates for major civil engineering projects are exceeded when the work is carried out. It is essential for the report to state clearly that the estimate is liable to considerable increases, particularly if the scheme is long delayed. It is wise to include in the report a forecast of increased costs based on a series of future dates for starting the work.

With regard to borrowing consents, local authorities have to obtain government consent to any scheme which is to be financed by loans. The departments concerned have to be convinced of the need for the scheme, that the design proposed is the most economical possible, and the cost/benefit return is satisfactory. The present government restriction on local-authority expenditure is severe, but fortunately the need for this has brought about the realization, both by government departments and local authorities, that positive forward programmes of capital investment in the various fields are essential.

Mention has already been made of changes to come in the structure of local government. We can only hope that the new system will, *inter alia*, have a radical effect on reducing the present major causes of delay between the inception and completion of capital projects. However, one is pessimistic in thinking about these aspects because many of the delays which occur are due to the 'democratic' principles which have to be observed. At every stage in local-government work, one has to bend over backwards to ensure that the citizens are given all possible facilities to object to what is proposed and to have their views noted and considered. In particular, the present practice of holding formal public inquiries into objections consumes a vast amount of time and money, and can result in schemes being delayed for months on end, while a number of relatively minor and often facetious objections are solemnly considered and debated. Streamlining of inquiry procedure is urgently needed. The present formal procedure should be abolished and replaced by an informal process, which would still be completely fair to all concerned. The public participation proposals in the Skeffington Report may, if sensibly applied, result in some reduction in objections.

PREPARATION

When the council has approved the scheme in principle, and a firm indication of loan sanction has been obtained, the engineer can proceed with preparing the detailed designs. The design of public works of any magnitude nowadays must be carried out by a team of specialists who constitute a 'working party'. The engineer must ensure that he is the controller and not allow any usurpation of his powers by other professions.

Above-ground surveys have been greatly assisted in recent years by the development of aerial photography, which has now reached the stage of complete reliability and accuracy. However, subsurface survey techniques are still far from reliable. Much more research is needed before a satisfactory answer will be achieved. The estimate of cost must include a generous provision for this work.

Before any design work is started, the engineer must decide whether it is to be done in his own office or by consultants. Sometimes a combination is the best answer, wherein the more straightforward work is done in the office, and a consultant or consultants are brought in to deal with certain specialized aspects. Far from there being any question of rivalry between consultants and local-authority engineers, it is probably true to say that we assist each other. Local authorities' offices provide an excellent training ground for young engineers, who get a wide experience. Many of these move later to consultants and to contractors. Then in the case of the more senior posts, there is or should be a corresponding movement back, thus allowing young engineers to obtain good experience in the three compartments.

Another major operation inevitably interlocked with design is fitting in the project with the requirements of the statutory undertakers (gas, water, electricity, telecommunications). Particularly in urban schemes for roads, tunnels, subways and sewers, the presence of mains belonging to the statutory undertakers is often the key factor in decisions on the feasibility of the scheme. In order to expedite the design and construction operations, coordination with the undertakers must be a highly sophisticated process, with regular meetings of the engineers concerned, and the securing of full cooperation and understanding of the design problems. During the execution of the works much time and cost may be saved if the main contractor is allowed to carry out the excavation work required by the undertakers, and even part, if not all of the work of laying the new pipes and cables. A liberal approach to the problem is essential. Much remains to be done in the way of convincing undertakers that well-qualified and equipped contractors can carry out all the work required by them.

During the execution of the work all legal requirements must be observed. Care must be taken to ensure that the necessary notices are served on private and public property owners. Complete records must be made of the physical condition of properties affected by the works. The employment of properly qualified surveyors is essential. Delays and risk of subsequent litigation can be minimized if properly skilled men are employed by the local authority to keep in close touch with their opposite numbers acting on behalf of property owners.

Obtaining consent of the local planning authority to the proposals is not always easy. Particularly is this so in the case of major urban road schemes where, for example, a new urban motorway inevitably has a great impact on the structure of the town and has side-effects, not only on the environment through which the road passes, but also on quite general aspects of the planning of the town. Not only must the detailed design be approved by all the local-authority committees concerned, but there are usually also external bodies who have to be consulted. For example, the Royal Fine Art Commissions in England and Scotland, as the case may be, have to be informed of the proposals for important schemes which are on or above the ground level. In particular, viaducts, flyovers, interchanges and bridges are of great interest to the Commissioners. The best possible architectural advice must be obtained at the start of the detailed design work, and an early informal approach made to the Commissions.

EXECUTION

The local-authority engineer has to advise on the appropriate method of executing the work. Often he is faced with political factors. Some local authorities are executing even large public works by the employment of direct labour. Others (for political reasons generally) favour the use of contractors for all but minor schemes. In theory it is possible for a local authority to set up an efficient direct-labour civil engineering organization with an adequately paid manager and staff, but there are many examples of schemes which have failed due to lack of efficient management.

If the work is carried out by contract, although the engineer is relieved entirely of responsibility for the organization of the works, he has to take on a heavy burden which starts with the preparation of the contract documents and ends only when the final account has been settled with the contractor. The present

British method of organizing contract works has been evolved over many decades, but is still a long way from a satisfactory *modus operandi*.

The 1964 Banwell Report on the *Placing and Management of Contracts for Building and Civil Engineering Work* made a number of recommendations for improvement, which were excellent in their intention, but lacking in really positive proposals. In fact, they contained a good deal of wishful thinking. In 1965 the Economic Development Committee on Civil Engineering set up a Working Party to consider implementation of the Banwell recommendations. It reported in 1968 and made some practical and positive recommendations. Once again, however, many of the more contentious problems were (quite rightly) left open pending further consideration. The conference on contracting in civil engineering since Banwell, at Solihull in June 1969, made a brave attempt to deal with a number of difficult matters in only one day. The conference partly achieved its objective of providing a forum for the frank expression of opinions both by contractors and professional engineers. As many of the points are of great importance some reference will now be made to them.

Site and soil investigations—The importance of adequate work was agreed by all. Strong disagreement was expressed on the question of the allocation of responsibility for subsoil surveys and their interpretation.

Selecting the contractor—The current practice in larger local-authority schemes seems to be to use selective tendering. This is working reasonably well in the writer's experience, but it is most important in these days with many contractors anxious to obtain jobs, that the number of tenders invited should be quite high. If the bill of quantities is kept short and simple, the expense involved by contractors in preparing tenders is not great in relation to their other standing overheads. Therefore, in the writer's opinion selective tendering should not be too selective.

Claims arising during the course of the work—The Banwell

Report had practically nothing to say on this matter; the 'Harris' Report dealt with it at some length, but did not have any positive answer to the basic problem. In this matter we are faced with a fundamental weakness in human nature, in that we do not seem to be able to trust each other. It is obvious that a completely new outlook is required on both sides. The representatives of contractors and engineers must get together on a national basis and endeavour to establish a code of conduct, which all concerned should solemnly undertake to observe meticulously.

Two major sources of claims are (1) extra costs caused by unforeseen sub-surface conditions; and (2) loss of productive time due to adverse weather. The classic type of claim arises when the contractor finds on excavating that the ground is different from that indicated by the subsoil survey. One way of reducing claims of this nature may be to bill the whole of the excavation work on a cost-plus and target basis. The traditional method, whereby the contractor has to assess the situation in advance, requires the engineer to do everything he can to specify, without ambiguity, the circumstances in which claims will be considered. Engineers must at all times realize that they are independent professional men, whose duty is to make the fairest possible assessment of contractors' claims. They must not think of themselves as acting solely for the client, and should not be biased by any thought of saving the client's money in cases where extra payments are clearly due.

With regard to (2), one relatively new method is to agree with the contractor the number of working days which will be allowed for the complete job instead of a fixed period for the work. On this basis the contractor has some security against loss of productivity due to bad weather, but there still remains the difficulty of agreeing on the degree of adverse weather which prevents working on any particular day.

Bills of quantities—Possible improvements on the traditional form of bills of quantities are now being examined by the Construction Industry Research and Information Association.

Their Research Project No. 98 is due to be completed early in 1971. Meantime their interesting interim report was published in the Proceedings of the Institution of Civil Engineers in January 1970, and is well worth study.

Many other important points were discussed at Solihull, such as pre-planning and programming, critical-path work, avoidance of delays in payments, serial contracting, negotiated contracts, package deals etc. However, space restrictions do not permit any further comment.

OPERATION

Once the scheme is completed and put into use, if it is well designed and constructed the end-product can hardly fail to function efficiently. However, this is not always the case, and the performance of a project is not always fully up to expectations. Engineers are remarkably reticent on these matters. Few papers or articles are published which deal frankly with these aspects. There is a need for more information to be made available. In the public interest engineers should not hesitate to prepare reports which, if considered too delicate for general publication, should be made available on a confidential basis to those who are concerned with designing similar schemes.

CONCLUSION

The writer hopes that this paper will have the desired effect of promoting interchange of views at the Symposium. The organization of capital projects is a vast and complicated subject. Only a few of the more important aspects can be dealt with in a paper of this length.

IO

SUMMING UP

Maurice Milne

There's a chiel among you takin' notes and, faith, he'll print them—Robert Burns

TO COMPRESS two days' discussion into small compass may do an injustice to the wide-ranging views expressed and inevitably it becomes one person's selection of what was important. Others would have chosen differently.

In the discussion on Mr. Erskine-Murray's paper we were taken through the implications for civil engineers of what management in his terms involved. It was clear that a successful civil engineer requires a wider competence than a narrow knowledge of the techniques of his profession involves. Such competence can be nurtured and indeed, having regard to the derivation of the word management, involving the concept of good husbandry, one can see that competence will grow given a suitable environment and encouragement.

The methods of achieving this are varied—sandwich courses, 'in-house' training and programmed learning—the last a useful and speedy method of acquiring knowledge of new techniques.

It is important that the senior executives are familiar with the subjects which junior staff are expected to assimilate on management courses, otherwise frustration will ensue when they endeavour to put their new-found knowledge into practice.

The Institution of Civil Engineers, like all other members of the CEI, includes in its requirements for corporate membership a paper on *The Engineer in Society*, a recognition that unless the engineer is aware of the impact of what he does on those around him he cannot regard himself as fully competent. The question arises, however, as to whether this goes far enough in promoting a study of management as part of the engineer's training—because training is now recognized as an important factor in developing management skills. 'Sitting alongside Nelly' is no longer good enough. To rely on acquiring knowledge of how men respond to demands made on them merely by being pitchforked into the working situation is not now adequate.

There is an enormous amount of new knowledge to be acquired and random reading has been suggested as one way to find out more of what is going on. But more disciplined study directed towards the needs of the particular organization by appreciation courses and 'in-house' training may well prove necessary to the young engineer.

When should this be done?—probably after corporate membership has been obtained. How?—by polytechnic and university courses, by programmed learning, by 'in-house' training and by one- or two-day symposia like the present one. Who should take such courses?—probably every engineer at various stages in his career would profit by just such opportunities.

The second speaker, Mr. R. L. Wilson, involved us in the human problems which arise whenever one deals with other people. What makes the worker want to work? Fear was the old way; paternalism succeeded it, but with increased mechanization we have had to think again and of new ways. Given the right environment there is no need to force a man to work—he will wish to do so because work is as natural as rest or play. Men will drive themselves to achieve their initial objectives, but for sustained effort with the commitment involved they must see their own needs being met.

To do this successfully means involving them in the organization within which they work. With professional men this means delegating to them real responsibility and authority to carry out their assigned task. It is surprising to find how widely imagination is distributed in the population and how little of this is tapped in modern industrial life. A successful manager will release this for the good of the organization and so arrange the way the work is distributed as to encourage acceptance of responsibility.

In design offices the current tendency is towards the formation of project teams and provided they are aware of the house rules with the trust this involves, the risk of their failing adequately to consult is small—but they must accept the need for this. Top management in whom final accountability reposes must have enough passing in front of their eyes to intervene when absolutely necessary. Delegation must be distinguished from abdication and one must not delegate to incompetents.

Interest in management begins when one is managed oneself and those who are being managed have a responsibility for training their manager. All too often the subordinate only tells his superior why he is leaving when he leaves rather than helping to put things right at an earlier stage.

Quite surprisingly to some, a workman can be entrusted to do technician's work; a junior engineer can do a senior's job and it should be the rule to push a job down to as low a level as possible by asking—'What is it that this man can do that his superior is doing now?'

Job evaluation and job descriptions are means necessarily adopted in large organizations to suit the man to the job or the job to the man and can induce rigidity which may conflict with the more free-wheeling approach referred to by Mr. Wilson. We must try to loosen up some of our attitudes to take advantage of the energy released by a less rigidly structured organization of the matrix type.

Trade unions may often inhibit such freedom by looking at the money and not at the quality of life in the work situation. But we are in an era where the quality of the environment in which we live concerns us and perhaps it is not too naive to look for a change in outlook, and indeed there are signs that this is so where people are no longer living at the subsistence level.

Young men seek responsibility early. Older men may see a threat in this, but there is evidence that the young will not give of their best if they see older men unjustly treated in an organization.

The third paper by Dr. G. S. Bosworth dealt with training for management and was introduced most amusingly. We were apprised of the difference between education, training and experience. Education is concerned with knowledge and understanding: training teaches us how to use knowledge to understand and solve problems, and experience arises from the real-life situation and brings confidence in the application of education and training. Understanding is achieved by perceiving how causes and effects in one system affect those systems which lie above and below it. For example, the use of the accelerator which affects the engine—a system below—and in turn speeds up the car in the traffic situation—a system above.

Training—the appreciating of problems—becomes important when one leaves the more sheltered learning situation and has to apply what one has learned. This is often a difficult transition. It is unethical in an examination to collaborate, but at work one must collaborate with others in a group to achieve the group's objective.

In so much of what we do in civil-engineering training we forget man himself either as an individual or as a corporate body. Indeed there are those who would argue that we ought to concentrate on the techniques of our profession and leave these other aspects to other people. This cannot be right.

How does an engineer become a manager? Certainly any successful engineer is a good manager. Design-and-make projects introduced into the training programme will teach in a simulative way how to collaborate with others in a team to achieve a desired objective. Harvard case-study system is another method. Sometimes schoolboys do better than graduates because they are less inhibited—a form of 'lateral thinking' perhaps? Such systems develop unorthodoxy and scepticism. They encourage people to discuss and argue—the 'think-tank' and 'brain-storming' sessions are similar in their aim. These methods have also shown the value of bringing academics into a works organization to the benefit of both. More cooperation between universities and industrial concerns should therefore be encouraged.

Young men in training should keep a notebook and jot down things that they have seen and learned and want to follow up. If these can be available to those responsible for training they are often revealing of the true situation as seen from the trainee's viewpoint and are a most salutary experience.

Command group training in management is a valuable technique where a whole vertical section of an organization from the general manager to the man on the shop floor come under a tutor for discussion and argument. A director may be asked 'What's your problem'—something he may never have been asked before. Individuals by these means are encouraged to develop themselves. They should be ready to accept responsibility for some of their own training rather than rely wholly on the firm.

The objectives of training need clear definition. Without these training can fail to satisfy the true needs of the individual or the job. The manager should ask—'What will I expect these graduates to be doing in five or six years' time?' and should make this clear to them.

When setting out job criteria one should have the ultimate aim clearly in mind. An example given was the job criteria for typists: obviously speed and accuracy are important attributes. But if a later examination of the typing pool shows a group of luscious blondes some other criteria were involved!

Discussion ranged over the value of business schools for management training. These have a value, but shorter courses of 'in-house' training may often have more practical value.

The following two papers by Messrs. J. Coats and J. Woodward on communication and personnel structure and Mr. Shennan's paper on staff selection were taken together.

As might be expected, communication was the theme. Peter Drucker's question was asked, 'Is there a sound in the forest when a tree crashes and no-one is there to hear it?' The recipient must be foremost in the mind of the communicator.

Thus a civil engineer must pay heed to the public he serves. He nearly always has to work in a team. He must get his ideas across to the client and the contractor, and all these require an ability to form meaningful relationships. To assist in this the analogy to a football team was drawn in an attempt to show this graphically. In pursuing the analogy in discussion the objective was lost in the technical discussion of how the team should be constituted and the relationships redefined. What was made clear, however, was how large a part of a senior executive's time is spent on finding out, in public relations, in giving instructions—all involving communication. Perhaps only twenty per cent of his time is spent in making decisions.

In the interview situation described by Mr. Shennan, this involves a two-way interchange. The man is being interviewed for the job, but he is also selecting the firm or rejecting it. Communication starts when the man comes into the room, the way he shakes hands, the way he sits down. The qualities being looked for in a man for site work are drive, understanding and ability to work in a team.

The selection of graduates may involve industrial psychologists and high-speed tests, but more usually the practice has been 'the milk round' of colleges and universities. These involve face-to-face contact with the men concerned, with the inherent difficulties of such confrontation for the young man meeting a senior member of the firm. This raised the question of written communication, the fear being expressed that this was choking and squeezing out example and the spoken word. A plea was made to reduce the torrent of paper—the curse of senior managements's life. The greater the elaboration on paper the greater was the danger of confusion.

H

The point was made that working drawings are the basis of communication between the engineer and the contractor, supported by specification and bills of quantities. The greater the elaboration of the latter the greater the tendency of contractors to claim.

Finally it was suggested that we were in a decade of public awareness after decades of technical advance. This meant that the ability to put our ideas across to the public would play a larger part in our work. Between professionals and others one of the best means of communication was a network which showed in graphical form the inter-relationship of activities without the need for a great deal of paper; where action was needed by whom and what would follow.

There followed a paper by Mr. Peter Dunican on management of the design process—design being concerned in all the decision-making which precedes the actual construction stage with all this implies. The credibility gap between design and building is considerable and can only be bridged by good communication—the most important activity in the whole process both in content and in timing. It is a two-way process including all parties involved who must communicate meaningfully. With the pressure from clients to get work started there is seldom sufficient time to tie up all the loose ends between the designer and the constructor. The customer should often be told to wait but seldom is.

The construction industry itself is under great pressure and works with all-too-small margins. When this happens the quality of service given suffers. When a job is losing money it may not be because of bad practice but because of external circumstances. What should be done? Some contractors withdraw into the laager and start shooting claims. The other way is to say 'we've bought this one' and try to tackle the situation dynamically, but still giving quality of service. This requires leadership of a very high order. The client must know the service he expects from you.

'The quality of management is strained'—a cry from the

heart. A plea was made for maintaining high standards of design. A contractor might yet say, 'I'm not prepared to erect this building—it's too ghastly for words!'

The fear was expressed that there is a danger in the present situation of the professional being displaced by the manager without training in any of the basic disciplines, bringing in the professional as he sees fit.

The Americans are adept at many of the technical skills involved in complex projects and get a better return on their resources by greater utilization. They make great use of models —mathematical as well as three-dimensional—and more often than we do indulge in multi-disciplinary practices. Management contractors may be employed by the main contractor to organize the work. Structural engineers are more often than civil engineers used to working in such teams in a subservient role.

This then led to discussion of the role of the quantity surveyor in civil-engineering work and contrast was drawn with the position in the architectural profession. The expanding role of the quantity surveyor often arose from the failure of the engineer to understand the broad economics of the work.

The discussion ranged on the role of the project engineer in a multi-disciplinary team. If this is too closely specified you can crucify the organization. On the other hand Mr. Dunican did not believe in the 'deep-end' theory unless you had a first-class rescue service. Wrong guys never work a perfect system!

On the use of computers much can be said. Here again we lag behind the United States. A Maryland firm has a terminal linked to a computer in Ottawa. Desk-top computers are in every office. The men see the value of them and trust themselves to use them. When a job fails they hope the computer will find the answer. The computer makes you think straight and not all engineers do this.

Reference was made to the RIBA 'Plan of Work', commending it to engineers. Finally the shortage of good detailers pointed to the use of a detailing service. This ran contrary to the ideas expressed by Mr. Wilson in his paper by breaking down the concept of a project team fully equipped with all services available directly to the project manager. There was also the problem of a satisfactory career structure for detailers.

Mr. Judson in his paper dealt with cost control in the design office pointing to the need for accurate estimating supported by adequate feedback. More money spent in planning and design may well result in less capital cost of the project, but forecast of design-costs can be difficult unless records are adequate.

To achieve this there must be a continuous monitoring process with continual review of the cost of the project as the design develops. The revised estimate leads to a revision of the commitment right up to the order to the contractor with an assessment of the work in progress against the budget. This is most necessary, for example, in the building of a power-station where designing and building go ahead concurrently and plant is being manufactured and costed as the building to house it is under way. To control such a situation requires a project-coding system with continual interchange between the designer and the construction side. As drawings are completed blocks of drawings are costed against the designer's coarser estimate and this requires very considerable skill. The effect of design changes must be costed, often requiring the knowledge of quantity surveyors outside the firm as well as within-house skills.

The study currently in train at Manchester University Institute of Science and Technology holds great promise—with the hope that simplification of bills will follow.

Money/time information can be stored in computers and speedy access to such information may help to improve the efficiency of estimating. The need for this is great with gains of 25–30% possible. For example on a power-station project the estimates were 30% out and the delay twenty-five months. The end of cost control is not at the end of the design process, but at the end of the job and in the collaboration in the early warning about a setttling of claims by contractors.

Mr. Rowbotham then dealt with management and cost-control of work on site. As he expressed it, 'management is the marshalling of men, materials and equipment against time and weather and human nature against cost of project to make a profit'. For site management, cost-control is a profit-preserving tool, and skill in its use takes time and experience. To secure the full benefits site management must assist in tender preparation, in assigning costs for work months or years ahead, and in assembling these costs so as to produce the lowest responsible tender. Secondly, site management must organize resources of men, equipment and use, and plan a schedule of work.

From an examination of these requirements it would appear that colleges need a new curriculum. There is a wide disparity at present in the efficiency of the industry and the failure rate is high. This may stem from the fact that site management has not the same interest in financial control as it has in other aspects of the works. The control of costs and spending must be done at the source—it does not just happen. There is a need for management accounting to assist in creating a policy for the organization. There follows a need for an examination of accounting techniques because there is no magic formula for maximizing profits. Such techniques could usefully be taught at some time in a civil engineer's career.

In the discussion the point was made that as soon as a man draws a line on a piece of paper a cost has been made. It behoves a designer, therefore, to be conscious of the cost of that which he designs.

On the whole there is an ignorance on site of cash flow especially when regard is had to the outstanding claims in the building industry over the country amounting to £600m. This, even if it were halved to £300m, is still a lot of idle money.

The division between the design side and the construction side of the industry may contribute to this lack of understanding and here the two papers—Mr. Judson's and Mr. Rowbotham's —taken together form a valuable link; but there is a long way to go.

Bills of quantities as now drafted may not be suitable for a proper cost plan; operational bills may be the answer. When a tender is prepared this is often done from first principles, while on the job it works in reverse. What is needed is a labour, plant and materials sheet, and only when these are properly recorded are costs properly controlled.

The final report of the Manchester study should be available soon based on the ICE conditions of contract. This is attempting to abstract the requirements of the industry and then to assemble experimental bills. It is hoped to reduce the money spent on measurement. There is a wealth of evidence to support the reduction in the number of bill items by 40%. If the bill of quantities can be so set up as to demonstrate the relationship between price and time it may be possible to get a realistic assessment of the proportion to put to quantities and the effect of time-related charges. It is most important to accept that all costs and all charges are not related to work completed. This could produce accelerated interim valuations. To deal with delay of claims going to arbitration, it is important to have a document to deal with the time aspect of the claim.

The ICE conditions of contract do not help with the major claim situation. If it were possible to get tenders to fix time-related charges this would assist in the rapid assessment of claims. If measurements could be simplified and varied, rates and claims quickly assessed it might bring down the £300m referred to which benefits nobody. Faults are not necessarily with consultants or clients—claims sometimes are made out at the beginning of a job and months go by without substantiation and only at the final account stage when losses are quantified are claims made out to redress the balance.

There is a need to pay promtly—on large contracts with costs running up to £$\frac{1}{2}$m per month, monthly interim valuations should be assesed and reconciliation made at three-monthly intervals.

The suggestion was made that the present scale of fees for consultants is adequate for routine work, but not for the management of complex work. Management contractors, as already suggested, are one possibility.

Turning to the 'cancer of claims'—these were separated into genuine claims which should be met promptly and other claims arising from contractors' failure to make a profit. Haphazard guesswork on rates due to inexperienced people with no feedback from site can cause this to arise. Estimators should be familiar with current site costs.

The comment was made that the agent allocated to a job was never consulted at tender stage. The difficulty is that site costs may not have a direct bearing on billed rates. Sites may be losing money because the firm charges high rates for the use of its own plant and as a whole makes a profit. Contractors fail to take agents into their financial confidence, and if a loss situation arises as described the agent may well try to retrieve it by hiring plant from outside at lower rates to minimize his own site costs.

Quality should not suffer, but contractors, site managers and resident engineers find that this can conflict with rigid cost control. Both sides must liaise when cost restrictions are pressing them to do nugatory work. When this occurs the right way to deal with it is cooperative effort to get the correct economic answer and leave to further discussion the contractual position.

Both sides should understand good site organization to make sure that what is reimbursable is measured. Cost control can decide beforehand what to do and what to cost and subsequent action merely confirms this.

Time is money and the cost of money varies month by month and so is not a constant factor. The time allowed to quote is often not sufficient properly to finalize the costs—and this may prevent site management from achieving its main function—to make a profit.

The last paper of the Symposium was that of Mr. F. R. Dinnis on the organization of capital works of a local authority. Mr. Dinnis took us through his paper suggesting that we in civil engineering were fearfully old-fashioned. He felt we had made little progress in some aspects of our work in the last fifty years. Listening to the discussion on previous papers he felt that our industry should pull its socks up and he particularly commended the study going forward at Manchester University Institute of Science and Technology on bills of quantities.

In local authorities the work was bedevilled by politics. With cost inflation going on all the time estimates had to take this into account. Delays arising from public inquiries and public participation—we were in the decade of public participation and much could be said in favour of it—cost money. The participation of non-elected bodies could devalue the role of the elected member.

The reluctance of the town clerk to risk going outside the law was a restraint that had to be accepted. We might have to contemplate the emergence of city managers—as had been tried in Newcastle.

He touched on the Banwell Report and its recommendations and the report of the 'After-Banwell' Committee under Sir William Harris—aimed at improving our tendering procedures. Direct labour was employed in local-authority work, but not to the exclusion of contractors.

He pleaded with his colleagues to be franker with each other not only in the feedback on costs but in 'what went wrong or what went right' on jobs. Even if this were circulated on a confidential basis it could be of enormous value. As for arbitration—avoid it!

The reference to participation and the role of amenity societies provoked a reaction from the floor. Such bodies prompted weak councils to action and acted as a local longstop, thereby performing a valuable service to the community.

One defect of our public inquiry system was to provide a forum for objectors, but seldom prompted the supporters of the scheme to come forward. The negative side of the case then received undue publicity.

The role of planners in the work of local authorities prompted the view that they had a place particularly in the planning and transport fields provided they had a sound engineering training. The independence of the professional man, employed though he be by a council, must be preserved. From the time of Telford onward the code of the civil engineer had been one of integrity and honesty from which a sense of partnership could develop. If a contractor and an engineer could trust each other this could reduce the incidence of claims. If ambiguity of documents were removed and more time for tendering and accurate pricing given, claims might be reduced and integrity and a sense of partnership on site develop. Is there a case for a code of conduct and could the Federation of Civil Engineering Contractors and the Institution look into this?

There was again a reference to discounted cash flow, which if properly applied to the examination of tenders might show that the lowest tender was not the most economical. Was there a hope that the finance committees of councils might try this? Mr. Dinnis felt that with computers to do this for us there was a glorious life ahead. If Skeffington were tried, it might ultimately reduce objections by better publicity for schemes.

One speaker reminded us that professional engineers do not receive a halo with their corporate membership. The pressure on the employed engineer was great. The role of the quantity surveyor was to save money for his client and if he were to play the dominant role with engineers as with architects this could undermine the engineer's position.

In engineering only one thing is certain—uncertainty. Thus, if it were possible to bring together consultants and contractors to devise a code of conduct, this could have the effect of promoting a feeling of mutual confidence between both sides of the industry. At one time contractors had few qualified engineers on their staff, but this is no longer true. Design engineers and contractors' engineers are together in our profession and improved communication between them should be for the better organization and management of civil engineering.

Regret was expressed at the absence from the Symposium

of representatives from the supply side of the industry who might
have had a valuable contribution to make to the discussion.

To sum up, there was a general feeling that the two days of
discussion in which so many had taken part and where discussion in almost every case had to be brought to a halt by the
passage of time, had been greatly stimulating and of very real
value to all. The hope was expressed that they provided a
fitting tribute to an engineer who had contributed so much to
the profession—Professor W. Fisher Cassie.

BIBLIOGRAPHY

Education and Training for Management
in the Construction Industry

A list of Books and Papers prepared by Dr. S. H. Wearne

Management Problems

BOWLEY, M. E. A. (1965) *The British Building Industry: Four Studies in Response and Resistance to Change*. Cambridge University Press.

BROOKS, H. (1964) Challenge to civil engineering. *J. Prof. Practice, Amer. Soc. civ. Engrs*, January 1964, 21–8.

CRICHTON, C. (1966) *Interdependence and Uncertainty. A Study of the Building Industry*. Tavistock Publications.

DARLING, T. Y. (1970) The contract system in civil engineering. *Civ. Engg Publ. Wks Rev.*, March 1970, 237–41.

ELSBY, W. L. (1967) A resident engineer's duties. *Proc. Instn civ. Engrs*, **37**, 411–14.

INSTITUTION OF CIVIL ENGINEERS (1969) Safety in civil engineering. *Proc. Instn civ. Engrs*, **44**, 143–52.

JOINT BUILDING GROUP (1967) Report of meetings held 1965–6. *Proc. Instn civ. Engrs*, **37**, 377–83.

MARSHALL, A. L. (1965) Management and the resident engineer. *Civ. Engg Publ. Wks Rev.*, June 1965, 831–5.

————— (1966) Rationalizing the approach to civil engineering management study. *Civ. Engg Publ. Wks Rev.*, February 1966, 191–4.

NATIONAL BOARD FOR PRICES AND INCOMES (1968) *Pay and Conditions in the Civil Engineering Industry*. Cmnd 3836. H.M. Stationery Office.

NATIONAL ECONOMIC DEVELOPMENT ORGANIZATION (1968) *Contracting in Civil Engineering since Banwell*. H.M. Stationery Office. (See also *Proc. Instn civ. Engrs*, **43**, (1969) 683–96.)

PARSONS, G. F. (1968) Duties of a contractor's agent. *Proc. Instn civ. Engrs*, **40**, 263–5.

THOMPSON, P. A. and others (1969) *Management of Public Works Projects*. Proceedings of a symposium. University of Manchester Institute of Science and Technology.

WEARNE, S. H. (1971) The contractual environment of management in civil engineering. *J. Management Studies* (in press).

Managerial Views

AUSTIN, W. T. F. (1968–9) Symposium on comparative project management. *Proc. Instn mech. Engrs*, **183**, part 3K.

HILL, A. J. (1964) Effect of formalized management training in the execution of civil engineering works. *Proc. Instn civ. Engrs*, **29**, 254.

NEWMAN, A. D. (1967) The importance of management. *Proc. Instn civ. Engrs*, **37**, 389–91.

PARSONS, E. A., O'HERLIHY, D. M. and ROWE, R. H. (1965) *Management in Civil Engineering*. E. & F. N. Spon.

TURNER, J. H. W. (1963) *Construction Management for Civil Engineers*. C. R. Books.

WARMISHAM, B. (1968) A system of incentives. *Surveyor*, 19 October 1968, 55–9.

Teaching Needs

CAMERON, A. M. (1964) Management courses: their timing and duration. *Proc. Instn civ. Engrs*, **27**, 862–7.

HOLBEIN, A. M. (1946–7) The training of foremen for civil engineering work. *Instn civ. Engrs Works Construction Div. Paper 7*.

INSTITUTION OF CIVIL ENGINEERS (1969) Conference on education and training, 1968. *Proc. Instn civ. Engrs*, **42**, 153–68, and **44**, 67–71.

LAING, J. M. and STRADLING, D. G. (1958) Management and management training in building and civil engineering. *Chartered Surveyor*, **91**, 151–4.

MARPLES, D. L. (1966) Management training for engineers. *Proc. Instn civ. Engrs*, **34**, 108–9.

PATEMAN, J. D. (1966) The formal education of the civil engineer. *Civ. Engg Publ. Wks Rev.*, July 1966, 839–42.

SOWDEN, A. M. and OVERSBY-POWELL, G. H. (1965) Management training from the point of view of the employee. *Proc. Instn civ. Engrs*, **31**, 442–5.

SYMINGTON, A. A. The training of inspectors for civil engineering work. *Instn civ. Engrs Works Construction Div. Paper 8*.

WYNNE-EDWARDS, R. M. (1965) Presidential address. *Proc. Instn civ. Engrs*, **30**, 1–16.

Research and Development

BARNES, N. M. L. (1970) Civil engineering bills of quantities. *Proc. Instn civ. Engrs*, **45**, 131–4, and **47**, 83–4.

MILLER, R. J. (1970) Lid off the smaller firm. *Consulting Engineer*, February 1970, 36.

WEARNE, S. H. (1969) The relation of management research to practice and to training. *Proc. Instn civ. Engrs*, **43**, 501–2.

Teaching Material

ABRAHAMSON, M. W. (1969) *Engineering Law and the ICE Contracts*. Second edition. McLaren & Sons.

ANTILL, J. M. and RYAN, P. W. S. (1968) *Civil Engineering Construction*. Third edition. Angus and Robertson.

BRECH, E. F. L. [Editor] (1970) *Construction Management: Principles and Practice*. Longmans.

CLOUGH, R. H. (1968) *Construction Contracting*. Second edition. John Wiley.

DRESSEL, G. (1968) *Organization and Management of a Construction Company*. C. R. Books.

FOSTER, C. (1969) *Building with men*. Tavistock Publications.

GILL, P. G. (1968) *Systems Management Techniques for Builders and Contractors*. McGraw-Hill.

INSTITUTION OF CIVIL ENGINEERS (1963) Civil engineering procedure. *Proc. Instn civ. Engrs*, **26**, N8–9.

————— (1969) *Introduction to Engineering Economics*. Institution of Civil Engineers.

JONES, R. L. and TRENTIN, H. G. (1968) *Management Controls for Professional Firms*.

MARKS, R. J., GRANT, A. and HELSON, P. W. (1965) *Aspects of Civil Engineering Procedure*. Pergamon Press.

MINISTRY OF PUBLIC BUILDING AND WORKS (1967) *Network Analysis in Construction Design*. H.M. Stationery Office.

MILLER, E. J. and RICE, A. K. *Systems of Organization*. Tavistock Publications.

NAVE, H. J. (1968) Construction personnel management. *J. Constn Div., Amer. Soc. civ. Engrs*, **94**, 95–105.

OPERATIONAL RESEARCH SOCIETY and INSTITUTE OF COST AND WORKS ACCOUNTANTS (1969) *Project Cost Control Using Networks*.

PILCHER, R. (1966) *Principles of Construction Management*. McGraw-Hill.

RUBEY, H. and MILNER, W. W. (1966) *Construction and Professional Management*. Collier-Macmillan.

THOMPSON, P. A. (1970) The cost of indecision. *Consulting Engineer*, March 1970, 44–7.

TUNG AU and others (1969) Papers on building construction games. *J. Constn Div., Amer. Soc. civ. Engrs*, July 1969, 1–38 and 85–106.

TURNER, G. J. and ELLIOTT, K. (1964) *Project Planning and Control in the Construction Industry*. Cassell.